Closed Loop
Electrohydraulic Systems Manual

Vickers Closed Loop Electrohydraulic Systems Manual

2nd Printing 1998
ISBN 0-9634162-1-9
Copyright © 1992 by Vickers, Incorporated
Training Center
2730 Research Drive
Rochester Hills, Michigan 48309-3570

Preface

This text is one of a future series of texts, targeted at the junior hydraulics technician who already is well versed in Basic Hydraulics Concepts and wishes to expand his knowledge in the area of electrohydraulic Closed Loop Control systems.

It is assumed that the reader has had at least some elementary training in Electronics, so that he/she is familiar with Voltage, Current, Resistance, Capacitance and Inductance in D.C. circuits, and the fundamental operation of amplifiers, ramp generators and comparators. This background information is encompassed by the first text, "Basic Electronics for Hydraulic Engineers." It is highly suggested that the first text be studied thoroughly before approaching the concepts in this manual.

Besides having good reading skills, the student must also be capable of using High School level algebra to understand the relationships between the various factors that cause Closed Loop Servo systems to react the way they do. However, the authors have strived to avoid the use of calculus, since higher mathematics seems to be unpopular at this level of instruction. A few basic calculus concepts are presented, in the hope that the student will develop an interest in this powerful technical tool.

The "Closed Loop Systems Training Manual" is the culmination of many hundreds of manhours of effort – the authors can take only partial credit for those hours. The text is based upon a technical paper by Vickers engineer Nalin Shah. Nalin researched and developed many of the formulas used in this text, with the goal of making the technology "reachable" by someone untrained in higher mathematics.

This is actually an Americanized expansion of an earlier, abbreviated version written in the United Kingdom by Stephen Skinner, Vickers International Training Manager – Europe and Asia. The earlier version played a key role in enabling our own internal training, and served as the basis for this text.

After a brief introduction in Chapter 1, the manual begins by defining the difference between an Open Loop and a Closed Loop in Chapter 2. The entire spectrum of electrohydraulic control valves is discussed, along with their intended areas of use. The three major applications areas (Position, Velocity and Force Control systems) are related to the best choice of valve type for each area.

The individual components of a Closed Loop system are discussed in Chapter 3. The operating characteristics of servo valves and high performance proportional valves are covered simply, and in considerable detail, in this chapter. Operating characteristics and specifications of that most crucial Closed Loop component, the feedback transducer, are also highlighted and explained.

Chapter 4 deals with proper sizing of control valves. The material presented in this chapter was developed in order to end the most common initial problem with proportional-type control valves; improper or poor operation caused by using the wrong valve size.

In electrohydraulics, one is presented with two major options. Either the actuator can be directly controlled, or the pump supplying power to the actuator can be controlled. Chapter 5 describes the advantages and disadvantages of these two options. The inescapable need for proper filtration in electrohydraulic systems is also driven home in Chapter 5.

Chapter 6 provides the analytical tools needed to estimate the predicted performance of a closed loop system. Gain, frequency response, stiffness and accuracy are the major concerns of this chapter. Since algebraic methods are used to calculate these parameters, real-world application of the methods described will require that an iterative process be used to find the optimum valve size and gain values.

Advanced control methods are explained in Chapter 7, where the concepts of Derivative and Integral gain are described in detail. Other control methods are also described. (It is assumed that the reader has not studied calculus.) Finally, Chapter 8 presents three actual applications, worked through to completion.

Digital control methods have not been included within this text. The intent is to eventually produce a separate text covering this subject, based on the assumption that the reader has had no prior digital electronics training.

Stephen C. Skinner, BSc   Richard J. Long, BSEE
International Training Mgr.   Technical Specialist
Vickers Systems, Ltd.   Vickers Technical Training Center
Havant, Hants, U.K.   Rochester Hills, MI, USA

Table of Contents

Introduction

A traditional benefit of hydraulic systems is their ability to control large amounts of hydraulic power while using relatively small components. Control of hydraulic power is now routinely achieved by the use of electronics. Electrohydraulics has been well proven for many years in the mobile, industrial and aerospace industries.

There are several advantages to an electrohydraulic system over other types of control:

- Hydraulic fluid flow transfers heat away from the components, and lubricates all moving parts constantly. This ensures reliable operation and long service life.
- Hydraulic actuators, whether linear or rotary, can be designed with wide speed ratings and high continuous-cycle ratings. Cylinders and motors can be stalled, reversed or operated intermittently without exhibiting the excessive wear or damage typical of electromechanical devices.
- Hydraulic actuators are not limited by magnetic saturation effects, as are electric motors. Since torque is proportional to pressure difference, very large torques can be delivered by small hydraulic devices.
- Hydraulic actuators are capable of higher stiffness, faster response and acceleration, and more efficient transfer of power and energy. This produces better gain, accuracy and bandwidth.

The combination of electronics with its accuracy and speed, combined with the "muscle" and efficiency of hydraulics, has developed into a technology ideally suited to closed loop control applications.

Basic Control Concepts

Open Loop vs. Closed Loop

What do we mean by "Closed Loop Control"?

It is important to define precisely what this term means. In a closed loop system, **the system's OUTPUT is the variable being controlled**.

The easiest way to understand what a closed loop system does is by comparing it to an OPEN LOOP system.

With open loop control, the system responds to an input signal (often called the COMMAND signal) and varies the output accordingly. But the system cannot automatically correct for any outside "disturbances" or changes that might occur.

For example, consider the control of an automobile's speed (Figure 2.1). This is an example of an OPEN LOOP SYSTEM.

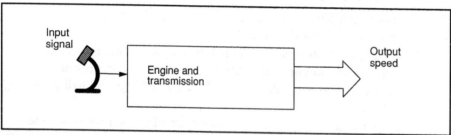

Figure 2.1

The input signal to control the speed of the car is the accelerator pedal. The further the pedal is pushed down, the faster the car goes. However, there is no accurate relationship between pedal position and vehicle speed. In reality, many other things will also affect the speed of the car, such as:

- The load the car carries.
- Head or tail winds.
- Travel uphill or downhill.
- Condition of the engine.

The system output, vehicle speed, is controlled by an input signal – pedal position. But external forces or system variations may affect the controlled output, as shown in Figure 2.2:

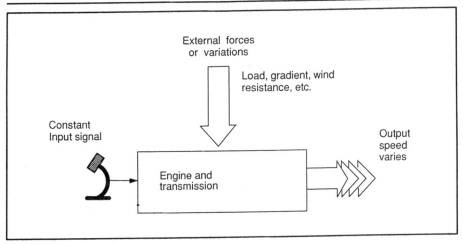

External forces
or variations

Load, gradient, wind
resistance, etc.

Constant
Input signal

Output
speed
varies

Engine and
transmission

Figure 2.2

Another example is the case of a hydraulic motor controlled by a simple throttle valve. The speed of the motor can be varied by adjusting the throttle setting.

But other factors will also affect the motor speed. These factors include:

- Load Pressure
- System Pressure
- Fluid Viscosity
- Leakage

Consider again the example of the car, but this time we'll outfit the car with a cruise control device.

Now, the required speed can be set by a potentiometer providing an electrical input to the system.

The vehicle is also fitted with a speed sensor which provides a FEEDBACK signal that is proportional to the actual speed of the car.

The cruise control device now has the ability to compare the desired speed (via the input signal) to the actual speed (via the FEEDBACK signal) and can automatically correct any difference between the two.

The signal from the speed sensor is called the FEEDBACK signal because it measures system output and feeds the information back to the system input for evaluation, as shown in Figure 2.3:

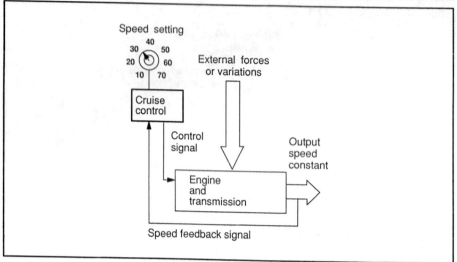

Figure 2.3

If the car started to climb a hill, the speed would tend to drop, creating a difference between input and feedback.

This difference would be amplified by the cruise control and sent on to the engine throttle as a control signal, which would cause the throttle to open and stay open until the actual speed equals the commanded speed.

The cruise control is an example of a CLOSED LOOP SYSTEM which maintains a constant vehicle speed regardless of outside forces. In control systems terminology, these outside forces are referred to as DISTURBANCE FORCES, because they "disturb" the system.

Many open loop systems are operated in a closed loop manner, but the error correction is performed by a human operator. This is normally what happens in, for example, a car that does NOT have a cruise control system. The driver decides the desired speed and pushes the accelerator to attain that speed. The actual speed is fed back to the driver by the speedometer. The driver then acts as the "cruise control system" and adjusts the throttle accordingly. However, the "system accuracy" is affected by the ability, the experience and the attentiveness of the human driver.

A true closed loop system senses the system output and performs the correction of any difference between input and feedback signals **automatically**, without human intervention.

The difference between input signal and feedback signal is called the ERROR. The goal of the control system is to keep this error signal as small as possible at all times.

Since the error correction occurs automatically, the system does not need to rely on the ability, senses, reaction time or experience level of a human operator.

Electrohydraulic applications of closed loop systems can be divided into three basic types:

Position Control   (Linear or Rotary)

Velocity Control   (Linear or Rotary)

Force Control   (Pressure, Torque or Load Control)

A system may apply a combination of types. A hydraulic press, for example, may be position controlled to a certain point, then switched over to pressure control to perform the actual pressing operation.

Some applications may also require the control of position and velocity at the same time.

Others may require sophisticated "motion profiles" involving very specific speeds, accelerations and positions in a complex and repeated pattern.

There are often many different ways of achieving the control objective. But, as the flexibility and accuracy requirements increase, so will the complexity and cost of the system. Therefore, it is important that a closed loop system be chosen to meet the system demands as closely as possible. This will ensure that the specifications can be met and that performance can be optimized, but at the same time ensuring that the system is not a costly "overkill" solution.

Position Control

The purpose of a position control system is to move a load to a certain position or series of positions.

The motion might be a purely <u>rotary movement</u>, as shown in Figure 2.4-A, where a missile guidance radar is turned by a hydraulic motor to point at a target.

The motion could be a purely <u>linear movement</u>, as in Figure 2.4-B, where a ram is punching holes in a sheet of steel.

In some cases, it may be desirable to convert rotary motion to linear motion as shown in Figure 2.4-C. The hydraulic motor turns a ball or lead screw to move the table.

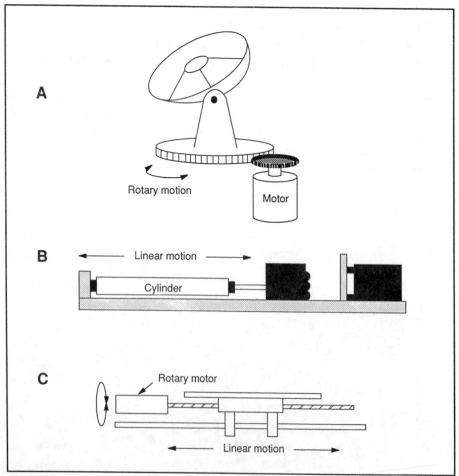

Figure 2.4

Ball screws, also called lead screws, are typically used when very high positional accuracies are needed.

One rotation of the motor will move the table a distance equal to the width of one thread on the ball screw. This "width" is called the PITCH.

If the pitch of the ball screw shown is .2 inches, then one rotation of the motor will move the table .2 inches.

One rotation of the motor equals 360 degrees. If we can control the rotation of the motor to + or − 1 degree, that means that we can control the motion of the table to within 1/360th of .2 inches.

The linear accuracy of the table motion would therefore be:

$$.2'' \times \frac{1}{360} = .00056''$$

In order to control the hydraulic actuator (i.e. cylinder or motor) in these applications, a valve is required which will, first of all, make the actuator move forward, move backward and stop.

If a sliding spool valve is used, we can represent this basic requirement with the three position valve shown in Figure 2.5:

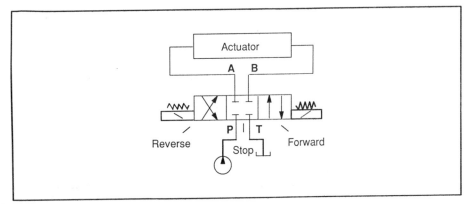

Figure 2.5

Assuming that the valve needs to be electrically operated, there are a range of valves to choose from, starting with simple On/Off solenoid valves at one end of the spectrum and extending through proportional and servo valves to microprocessor-driven valves at the other end, as shown in Figure 2.6.

Which one chosen depends on our accuracy requirements and the level of sophistication needed for the job.

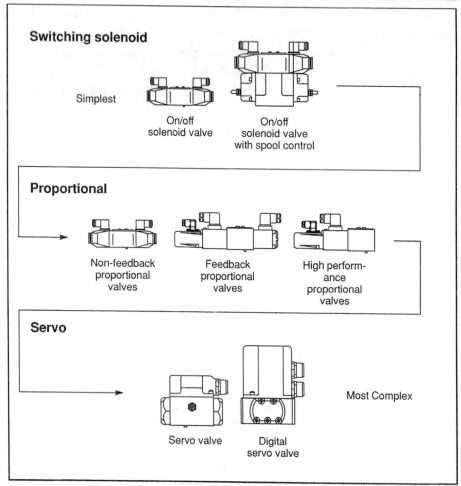

Switching solenoid

Simplest

On/off
solenoid valve

On/off
solenoid valve
with spool control

Proportional

Non-feedback
proportional
valves

Feedback
proportional
valves

High perform-
ance
proportional
valves

Servo

Servo valve

Digital
servo valve

Most Complex

Figure 2.6

On/Off Solenoid Valves

On/Off solenoid valves can be two or three position valves.

They may be either direct solenoid operated (solenoid directly shifts the spool) or they may be solenoid pilot operated (solenoid shifts the spool in the pilot stage, porting pressure to the main stage spool).

In either case, the flow of oil to the actuator is switched ON and OFF, and the DIRECTION of oil flow is also controlled.

This type of valve, on its own, does not provide any control over the actuator's speed or acceleration/deceleration (Figure 2.7):

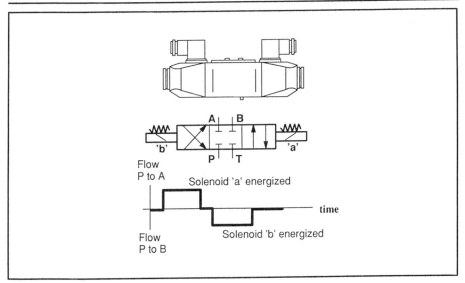

Figure 2.7

When solenoid 'a' is energized the spool shifts fully to the left, porting P to A and B to T.

When solenoid 'b' is energized the spool shifts fully to the right, porting P to B and A to T.

Flow in both directions is either fully ON or fully OFF.

Solenoid valves can be operated by a wide range of supply voltages, can be either AC or DC type, and can be operated by simple relays, switches or solid state devices.

Depending on the size of the valve, typical RESPONSE TIMES (the time required for the spool to move from one position to another) can range from 20 to 100 milliseconds.

A typical position control system using an on/off solenoid valve is shown in Figure 2.8:

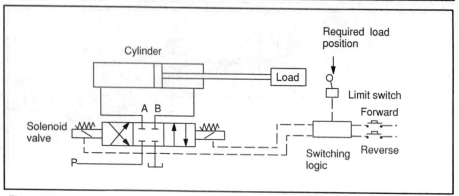

Figure 2.8

In this example, the valve is controlled by switching logic, which could be in the form of relay circuits, solid state switches or a programmable logic controller (PLC).

The cylinder can now be moved from its rest position to some other position determined by the location of a limit switch.

In this simple case, pressing the "forward" button would cause the cylinder to advance until the limit switch is operated. The cylinder would then stop at the required position.

The advantages of such a system are SIMPLICITY, LOW COST and, in some applications, an ADEQUATE level of performance.

However, there are a number of factors that will determine the exact position where the load will stop, and how REPEATABLE this motion is.

When the load hits the limit switch and activates it, the signal is sent to stop the load – but this CANNOT HAPPEN INSTANTLY.

The stopping distance, from the time the switch is actuated to the time the cylinder actually stops, will be affected by many things:

- The response time of the limit switch
  PLUS
  The switching time of relays used
  or
  The cycle time of the PLC (if used)
  PLUS
  The response time of the solenoid valve...
  ... all add up to delay the actual closure of the valve. The switch takes some measurable amount of time to close. Relays or a PLC will also take time to react to the switch signal, once received. The valve will then take from 0.020 to 0.1 second more, depending on

its size, to respond to the signal from the switching logic. The load continues to move during this time.

- The velocity of the cylinder

  and

  The mass of the load

  and

  The amount of friction acting on the load...

  ... all affect the load's INERTIA. Inertia dictates the amount of force that it will take to stop the load. If we move the load faster or make the load heavier, the load's INERTIA increases, making it harder to stop.

- The compressibility of the fluid

  and

  The internal leakage in the hydraulic components

  and

  The amount of oil under compression in the components...

  ... all affect the ability of the system to brake the moving load.

Some of these factors are difficult to determine accurately, and in reality, ALL will change during operation of the machine.

- The mechanical, electrical and hydraulic components will react faster as they "wear in" and then slow down as they "wear out."
- The viscosity of the oil will change with temperature.
- Compressibility of the oil will change as air molecules become entrained in it.
- Internal leakage in hydraulic components will tend to gradually increase with wear.

These factors result in an amount of uncertainty in predicting the exact position at which the load will stop (Figure 2.9):

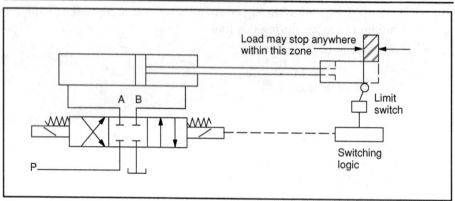

Figure 2.9

As you can see, the accuracy and repeatability of such an arrangement is limited.

There are other problems associated with this type of control. If the load is heavy or traveling at a high velocity, the sudden stopping of the load causes high pressure peaks and violent shock in the system. This results in leakage, frequent damage to components and dramatically reduced system life.

The system also lacks flexibility in that, if the stop position of the load needs to be CHANGED, you must physically MOVE THE LIMIT SWITCH.

If several different stop positions are needed, there must be a limit switch and the associated switching logic for EACH STOP POSITION.

This type of control is called a DISCONTINUOUS (sometimes referred to as a "Bang-Bang") closed loop system, because only an on/off feedback signal is provided.

Another example of a discontinuous closed loop system is provided in Figure 2.10, where a pressure switch is used to unload a pump in an accumulator system. In this instance, a discontinuous system is used to provide very adequate control for the application.

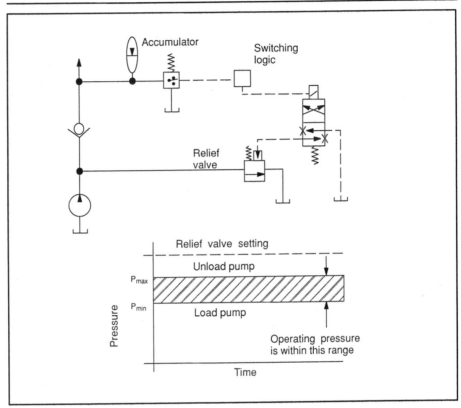

Relief valve setting

Unload pump

P_{max}

P_{min}

Load pump

Pressure

Operating pressure
is within this range

Time

Figure 2.10

The pressure switch controls the system pressure by providing an on/off signal to the unloading valve for the pump.

A certain amount of DEADBAND must be allowed in the pressure switch to prevent the pump from being continually loaded and unloaded.

This does limit the accuracy to which the system pressure can be maintained. At any given instant, the system pressure will be anywhere between P_{max} and P_{min}.

Perhaps the most familiar discontinuous system to you is the thermostat on your home heating system. This system uses a temperature switch to turn on your furnace automatically. The switch turns your furnace on when the temperature gets too cold, and turns it off again when the temperature setting has been reached. It also has a deadband of several degrees designed into it, to prevent the furnace from continually turning on and off. When operating properly, the temperature in your home is always somewhere between T_{max} and T_{min}.

The next valve in the spectrum is a conventional on/off solenoid valve that has been modified to provide a certain amount of control over the spool. There are several simple ways to do this.

STROKE ADJUSTERS:

In two-stage valves, stroke adjusters can be fitted to the body of the valve at the main spool. The stroke adjuster is simply a device which limits the amount of travel allowed to the spool.

This provides the valve with a degree of flow control by limiting the amount of opening between spool and port. However, this form of flow control is not compensated for pressure or temperature changes, and has to be adjusted manually.

PILOT CHOKES:

Two-stage valves can also be fitted with a block containing a pilot choke. This is simply a block containing orifices that restricts the flow of oil to the ends of the main spool, allowing the speed of motion of the main spool to be controlled. This allows a degree of acceleration and deceleration control over the actuator.

ORIFICE PLUGS:

Direct-operated solenoid valves can sometimes be fitted with an orifice plug to slow down the speed of spool movement (Figure 2.11):

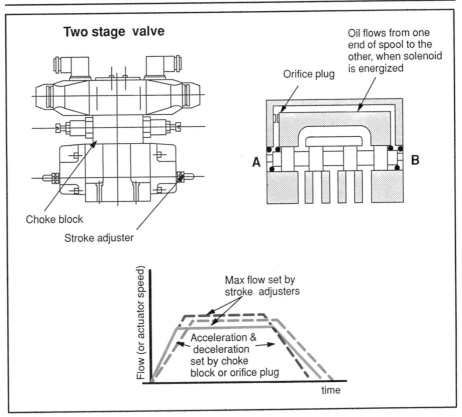

Figure 2.11

The application of these valves is basically the same as for simple on/off valves except that:

- It may be possible to control actuator speed within the valve itself.
- System shocks may be reduced by slowing down the spool movement, thereby regulating the acceleration and deceleration of the load.

Although restricting the speed of spool movement will help reduce shocks caused by starting and stopping the load, it is also likely to reduce the position accuracy of the system, since there is now one more factor to affect the system (Figure 2.12):

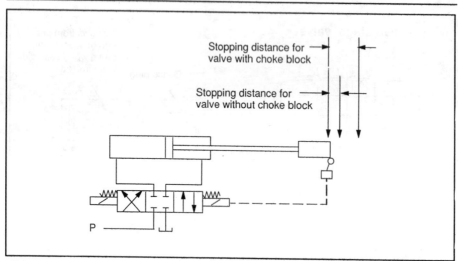

Figure 2.12

Variations in pilot drain pressure and fluid viscosity (due to temperature) will likely affect the exact point at which the load will come to rest.

Non-feedback Proportional Valves

This category includes both direct acting proportional valves without spool position feedback (no LVDT on the spool) and two-stage proportional valves with single stage feedback.

Proportional valves differ from on/off solenoid valves in that it is possible to control the spool position to any point within its range of travel by varying the electrical current to the solenoid.

This means that we can control the flow rate through the valve, and therefore control the speed of the actuator, by electrical means.

The schematic symbol for a direct acting proportional valve is shown in Figure 2.13:

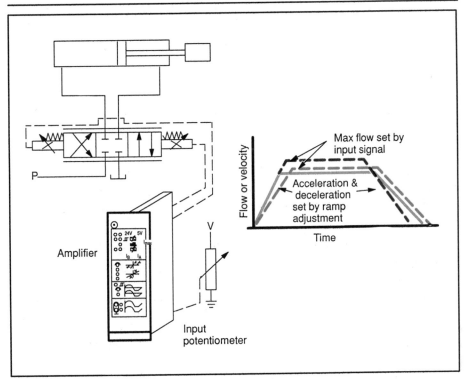

Figure 2.13

Note that the proportional valve symbol has a bar across the top and bottom of the spool, which indicates the spool's variable positioning in the valve body.

The solenoid symbols have arrows indicating their need for a variable current.

The current for the solenoid could be supplied directly from a potentiometer, making for a very simple circuit.

However, the potentiometer would have to be very large, in order to dissipate the heat that it would generate while supplying power to the solenoids.

Normally the valve will be driven by a control amplifier.

The amplifier is designed to accept a low power signal from a small input device (such as a potentiometer), amplify it to the necessary power level and transmit it as a drive signal to the solenoids.

But a well designed amplifier does much more than simply amplify the input signal. It will also:
- Minimize power lost through unnecessary electrical heat generation.

- Automatically compensate for changing solenoid resistance as the solenoids change temperature. (Solenoid resistance can change as much as 40% during operation.)
- Eliminate the effects of supply voltage fluctuations.
- Provide various adjustments to optimize the valve's performance, such as gain, deadband compensation, and ramping capabilities.
- Center the valve for safety purposes if power is lost or if a feedback device fails.

Typical response times for this class of valve range from 50 to 150 milliseconds, depending on valve size. The valve's response time can be considerably lengthened (up to several seconds) by using a simple ramp generator.

The ramp generator can be used to convert an on/off signal to one that gradually rises in a controlled manner as shown in Figure 2.14:

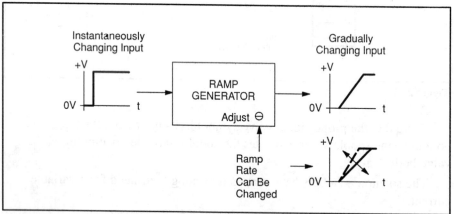

Figure 2.14

With a ramp generator, it is now possible to control the speed of spool movement, and therefore control the acceleration and deceleration of the actuator.

Since the spool movement is controlled electrically instead of hydraulically, spool motion is not affected significantly by changes in fluid viscosity or variations in pilot pressure.

In many applications, proportional valves may be used with simple limit switches to position an actuator in the same way as on/off solenoid valves. Their main advantage over the on/off valve, however, is that they can precisely control the acceleration and deceleration of the load, providing smooth mechanical movement and reducing or eliminating destructive system shocks.

Proportional valves basically provide non-compensated flow control.

However, non-feedback single stage proportional valves DO provide a certain amount of PRESSURE compensation, due to the design of their spools.

If an application requires more pressure compensation, a hydrostat can be used with the valve (Figure 2.15) to provide a constant actuator speed regardless of system pressure or load pressure changes.

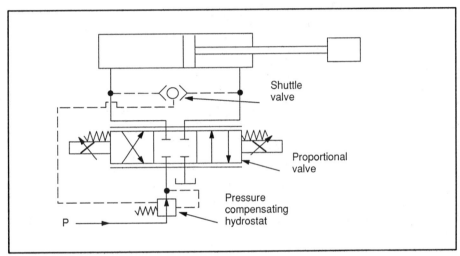

Figure 2.15

When a proportional valve is used with limit switches, it is being used in a discontinuous closed loop system.

It is also possible to fit the actuator or the load itself with a position transducer to achieve true, continuous closed loop position control.

Unlike a simple switch, a position transducer will provide a feedback signal which is proportional to actuator position – not just an on/off signal.

One very simple position transducer is a linear potentiometer, used in Figure 2.16:

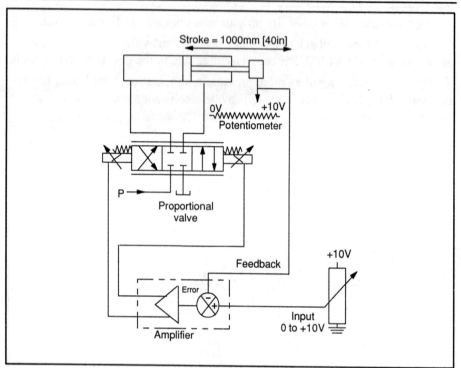

Figure 2.16

An input signal corresponding to the desired actuator position is fed into the amplifier.

The signal is amplified by the amplifier and sent to the valve solenoid, which moves the spool and creates flow through the valve to the cylinder.

As the cylinder moves the load, a feedback voltage is sent from the feedback potentiometer back to the amplifier.

This feedback signal enters the amplifier at the SUMMING JUNCTION. The summing junction subtracts the feedback signal from the input signal to produce the ERROR signal.

At any instant in time, the ERROR signal is proportional to the INPUT COMMAND SIGNAL minus the FEEDBACK SIGNAL.

Consider our example in Figure 2.16 again, where a 40 inch stroke cylinder is being used to position a load.

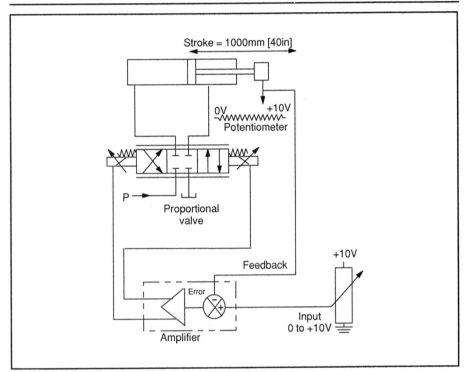

Figure 2.16

The input signal ranges from 0 to +10 volts, where

0 volts = Fully Retracted (0 inch position)
+10 volts = Fully Extended (40 inch position)

Therefore, each 1/4 volt applied to the input represents 1 inch of desired movement.

To simplify this example, the feedback pot is powered from the same +10v supply as the input command device, so the feedback signal also ranges from 0 to +10 volts, where

0 volts = Fully Retracted (0 inch position)

+10 volts = Fully Extended (40 inch position)

so each 1/4 volt of feedback signal also corresponds to 1 inch of movement.

The feedback signal is connected to an INVERTING input on the amplifier. This serves to invert the feedback signal (reverse its polarity from PLUS to MINUS voltage). This is done so that the SUMMING JUNCTION can subtract feedback from command by summing a positive command with a negative feedback. (NOTE: A NON-INVERTING input could be used for the feedback connection if 0 to –10 volts were available for the feedback potentiometer.)

Assume that the cylinder starts at the 0 inch (fully retracted) position.

Next, we apply an input signal in the form of a +5 volt step (Figure 2.17), by closing a switch. This 5 volt signal corresponds to a desired movement to the 20 inch position.

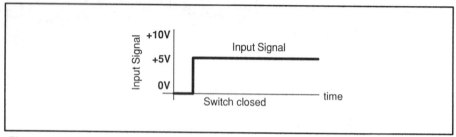

Figure 2.17

This step input will cause the valve to open and the load to move. As the load moves, the feedback signal starts at 0 volts and constantly increases.

Rather than try to comprehend the effects of the constantly changing feedback signal at this point, let's FREEZE the changes in one-second steps to simplify what is happening.

A fraction of a second after the +5 volt input is applied, the cylinder has not yet responded. This means that the input is +5V and the feedback is 0V.

The error signal will therefore be +5V , causing the amplifier to produce a corresponding output to the valve solenoid. The spool shifts, allowing flow through the valve.

The actual rate of flow (and therefore the speed of cylinder motion) will now be determined by several factors:
- The gain of the amplifier.
  (How much current will the amplifier output due to the +5V error signal at its input?)
- The flow rating of the valve.
  (How much oil can pass through the valve for the given amount of spool shift ? How big is the valve?)
- Valve pressure drop.
  (How hard is the pump pushing oil through the valve?)

Let's assume now that we have taken all of these things into account, and we have found that an input signal of +5V produces:

Cylinder Speed = 8 inches per second

If the cylinder continued to move at this speed for the whole 20 inches of motion, it would take:

$$\frac{20 \ \ inches}{8 \ \ inches/\sec} \ = \ 2.5 \sec$$

Consider, though, what would happen after 1 second. If the cylinder maintained its 8 in./sec speed, after 1 second it would move 8 inches.

After moving 8 inches, the feedback signal would now be:

$$8 \ \ inches \times .25 \ \ volts/inch \ = \ 2 \ \ volts$$

The error signal into the amplifier would then change to:

$$5 \ \ volts \ - \ 2 \ \ volts \ = \ 3 \ \ volts$$

If you assume that there is a direct relationship between input signal and flow through the valve, then the flow (and therefore the cylinder's speed) will now be 3/5ths of what it was when we started. The speed is now:

$$8 \ \ inches/second \times \frac{3 \ \ volts}{5 \ \ volts} \ = \ 4.8 \ \ inches/second$$

The speed has dropped from 8 to 4.8 inches/sec because the error signal has decreased.

During the next second, the cylinder would travel only 4.8 inches, for a total movement of:

$$8 \ \ inches \ + \ 4.8 \ \ inches \ = \ 12.8 \ \ inches$$

The feedback signal would now be:

$$12.8 \ \ inches \times .25 \ \ volts/inch \ = \ 3.2 \ \ volts$$

The error signal would change again, to:

$$5 \ \ volts \ - \ 3.2 \ \ volts \ = \ 1.8 \ \ volts$$

And the speed would decrease again to:

$$8 \ \ inches/second \times \frac{1.8 \ \ volts}{5 \ \ volts} \ = \ 2.88 \ \ inches/second$$

As you can see, the cylinder starts at maximum speed but continually slows down as it gets closer its commanded position.

The cylinder speed is determined by flow rate through the valve. The flow rate is determined by amount of spool movement in the valve. Amount of spool movement is determined by the output of the amplifier. Amplifier output is determined by the size of the error signal coming in.

Therefore, as the error gets smaller, the cylinder slows down.

Continuing the calculations would result in an output movement as shown in Figure 2.18:

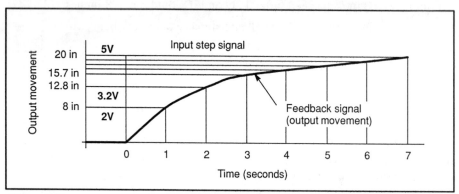

Figure 2.18

As you can see, our simplified system would take considerably longer than the 2.5 seconds that we calculated for the motion with constant application of maximum flow.

And in reality, our system would operate even SLOWER because it reacts to the decreasing error CONTINUOUSLY – not in one second increments.

This system would also produce very poor accuracy, because the error signal would have to be relatively large to create a significant output from the amplifier.

THE ANSWER TO THE PROBLEM is to turn up the amplifier's Gain, so that we keep the valve fully open through most of the cylinder's travel **until the cylinder gets close to its commanded position and the error signal gets relatively small.**

This is covered in greater detail in Chapter 6.

The response time and accuracy that you can achieve with a closed loop position control system is determined mainly by the characteristics of the actuator/load combination and control valve itself.

Direct acting proportional valves without feedback, and single feedback two-stage proportional valves tend to be rather slow, and can limit the performance capabilities of a system when response time and accuracy are critical factors.

A second limitation with the non-feedback proportional valve in position control applications is the fact that the spool has a large overlap in the center position (Figure 2.19):

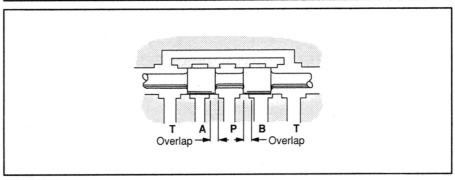

Figure 2.19

The edges of the spool lands overlap the ports in the valve body, creating a certain amount of "dead" movement of the spool. Flow through the valve does not begin until the spool has moved beyond this overlap region or **DEADBAND**.

This deadband region is designed into this type of valve for a couple of reasons:

First of all, the deadband region provides a degree of safety in the event of power failure to the valve. If power fails, the spool will spring to the center position and the comparatively large overlap ensures that flow through the valve is fully cut off.

The actuator will stop.

Secondly, the overlap also reduces the cost of manufacturing the valve because the spool lands do not have to be "matched" to the valve body ports with high accuracy.

Thus, this type of proportional valve is relatively inexpensive.

The effect of spool overlap on the flow characteristics of the valve is shown in Figure 2.20 below.

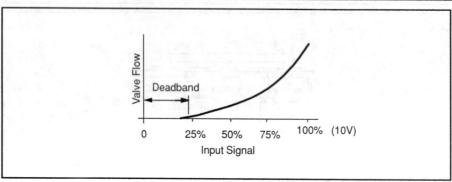

Figure 2.20

Flow through the valve does not start until the signal to the solenoid reaches approximately 25% of its maximum value, where the spool begins to move past the deadband.

If we were to look at the effects of deadband on our simple example of Figure 2.17, it would mean that whenever the error signal fell below 2.5 volts (25% of the 10 volt maximum input), flow through the valve would stop.

This would cause the actuator to stop after only 10 inches of stroke out of the 20 inches we wanted! (Figure 2.21):

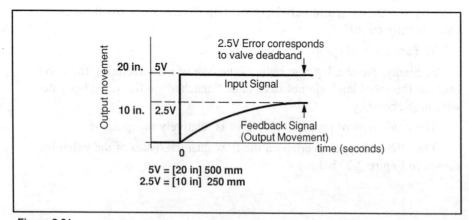

Figure 2.21

Our oversimplified system produces an exaggerated example of the problem. However, we could reduce this problem by increasing the gain of the amplifier. The error caused by the deadband would be reduced, but not eliminated.

In a real system, however, we can virtually eliminate the effects of deadband electronically, by adding DEADBAND COMPENSATION into the standard amplifier for the valve.

Deadband Compensation circuits automatically increase the gain of the amplifier when the spool is near the central region of the valve. This allows even a small error signal to move the spool across to the edge of the deadband.

This virtually eliminates the deadband by electronic means, producing the improved characteristic shown in Figure 2.22:

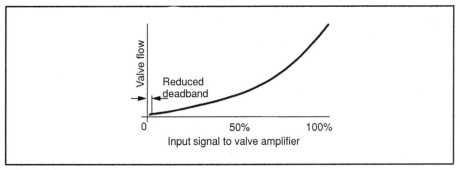

Figure 2.22

The deadband can be reduced from 25% of maximum input to about 1%.

Another factor which enters into consideration is HYSTERESIS. Sometimes called static friction, or "stiction," hysteresis basically tends to cause a difference in the amount of solenoid current necessary for identical spool displacement when the cylinder is EXTENDING to a commanded position as compared to when the cylinder is RETRACTING to that same position.

The hysteresis of the non-feedback variety of proportional valve tends to be relatively high (2 to 8%). Although its effects can be decreased through incorporation of a DITHER in the control amplifier, it will still contribute to the uncertainty in a positioning application.

Dither is a high frequency (60 to 100 Hz) signal which is constantly sent to the valve solenoid by the amplifier. It causes the spool to vibrate, in an attempt to prevent the spool from settling into position and encountering "stiction."

The bottom line, however, is that non-feedback proportional valves are not ideally suited to closed loop position control applications unless the system involved is a relatively slow moving system where high accuracy is not a prime requirement.

Feedback Proportional Valves

This group of valves includes both the direct acting valves with spool feedback, and the two-stage valves with feedback on both spools.

They have the same basic characteristics as the non-feedback valves, although their performance levels are higher.

	NON-FEEDBACK VALVES	FEEDBACK VALVES
Response Time	50 to 150 milliseconds	12 to 37 milliseconds
Hysteresis	2% to 8%	1%
Deadband	25%	25%

The effect of spool deadband will be the same as for non-feedback proportional valves, and therefore limit the accuracy of a closed loop position control system.

The faster valve response means that a faster system response can sometimes be achieved, depending on the characteristics of the actuator and load.

For certain applications, the possibility exists to produce special spools which have a reduced amount of overlap, allowing the valves to be used for the less demanding closed loop position control applications. This can provide significant cost and serviceability advantages over servo valves.

To achieve the required system response rate and accuracy, it may be necessary to incorporate an additional voltage amplifier prior to the standard valve drive amplifier (Figure 2.23). This increases the overall gain of the system.

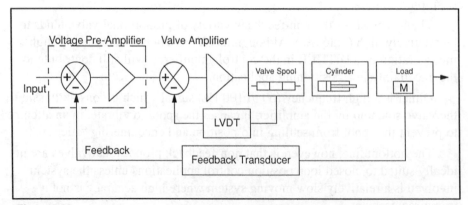

Figure 2.23

A common misconception surrounding feedback proportional valves is that they are only used in closed loop systems, as shown in Figure 2.23 above.

This misconception occurs because the term "feedback" is usually associated with a closed loop system.

It is important to note that, in this type of valve, the term "feedback" refers to the **spool position** feedback from its built-in LVDT. The position of the spool is fed back to the amplifier, to verify that the valve spool was moved by the commanded amount. This does NOT produce a true closed loop system overall.

This arrangement is frequently referred to as a **CLOSED INNER LOOP**. The system that the valve is controlling could still be either an open loop system or a closed loop system.

A system can truly be called a closed loop system ONLY if the feedback comes from the actual load or other system parameter being controlled, as shown in Figure 2.23.

Feedback proportional valves are frequently used in open loop systems, to take advantage of their better response time and lower hysteresis characteristics.

High Performance Proportional Valves

High performance proportional valves use a single solenoid to position the valve spool in the body (Figure 2.24):

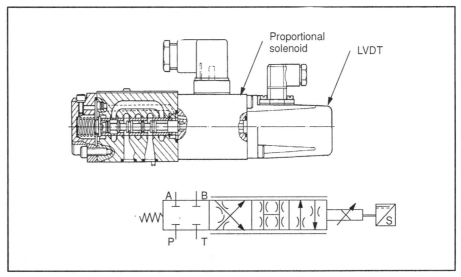

Figure 2.24

A position sensor (LVDT) creates a feedback signal which is proportional to the spool position in the valve. This feedback signal is connected to the valve's control amplifier.

When the solenoid is de-energized, the spool spring pushes the spool completely over to the all-ports-blocked position.

Therefore, the null position is achieved by partially energizing the solenoid, and the spool is moved to either side of null by increasing or decreasing the solenoid signal.

The design of the solenoid and LVDT assembly, combined with the special control amplifier which maintains null position, produces a valve with a fast response time (typically 10 milliseconds to respond to a 100% input step signal).

By using a sleeved body, the spool lands can be matched with the port openings very accurately, producing a valve with virtually no spool overlap.

The combination of fast physical response and zero deadband makes the valve suitable for many applications that could formerly be handled only by a servo valve.

Servo Valves

Servo valves are typically two-stage, or even three-stage, valves.

They commonly use a flapper/nozzle type pilot stage, as shown in Figure 2.25:

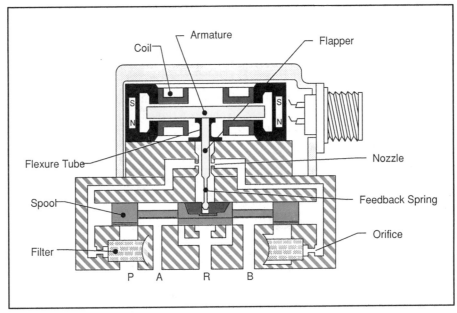

Figure 2.25

The pilot stage is operated by the torque motor and its attached flapper/nozzle assembly.

This positions the main valve spool on either side of the centered, or null, position depending on the direction and amount of electrical current flow through the torque motor coils.

The main spool slides inside of a sleeve which has precision-machined ports, allowing for accurate matching of the spool lands with the valve ports. This also enables a choice of overlapped, underlapped or zero lapped spools to be specified.

Servo valves have classically been the automatic choice for all but the simplest closed loop position control systems.

They offer very fast response (usually less than 10 milliseconds step response), very good linearity (input signal vs. flow) and low hysteresis.

However, use of a servo valve in a system requires good housekeeping on the hydraulic fluid. Adequate fluid filtration is vital, since a servo valve *will* malfunction if dirt enters the pilot stage.

In addition, state-of-the-art servo valves are generally not completely field repairable, and require special skills to disassemble and repair them correctly.

But where fast response, high speed and precise motion are required, they continue to be the primary solution for the most demanding applications.

Digitally Controlled Servo Valves

Servo valves are now available with a microprocessor built into the valve itself, allowing the advantages of digital control methods to be utilized. The hydraulic part of the valve is a conventional flapper/nozzle servo valve.

The torque motor is now controlled by a built-in microprocessor rather than by an analog amplifier. Figure 2.26 shows a typical application of the valve mounted on a hydraulic cylinder.

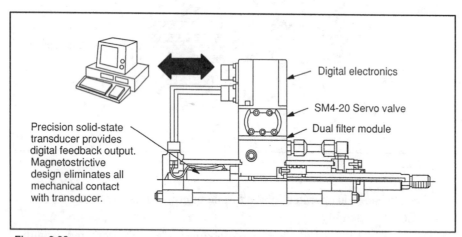

Figure 2.26

The cylinder has a position transducer built into it, providing a digital feedback signal to the microprocessor controller in the valve. The controller includes a feedback signal conditioner, a closed loop controller and an output drive amplifier for the valve's torque motor. Instructions are fed to the valve from either a personal computer (PC) or a programmable logic controller (PLC). Because each valve has its own address code, up to sixteen valves can be controlled from each output port of a PC or PLC.

A servo control algorithm is provided in the microprocessor's software which takes into account the various control parameters, such as load mass and valve characteristics. Additional software can be added to the host PC or PLC, or into a motion controller, providing a highly flexible multi-axis control system.

Velocity Control

Velocity control systems are used when it is necessary to control the speed of actuator movement rather than the position. This can be used with either a linear actuator or a rotary actuator.

In either case, we are actually controlling the FLOW RATE of fluid to the actuator.

There are many different ways in which actuator velocity can be controlled, ranging from a simple orifice in a hydraulic fitting to a closed loop servo valve arrangement.

The following sections describe the options available, and the characteristics of each option.

Fixed Orifices

The simplest way to control actuator speed is to use a fixed orifice installed in the port of an actuator, or in the line going to the actuator as shown in Figure 2.27:

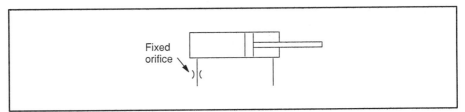

Figure 2.27

The orifice will restrict fluid flow in both directions and therefore will control speed when either extending or retracting.

If different speeds are needed for extension and retraction, two different restrictions could be used with bypassing check valves as shown in Figure 2.28:

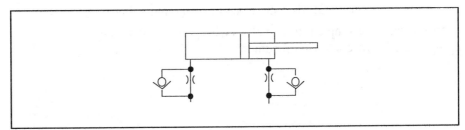

Figure 2.28

In either case, the orifice will not be the only factor which decides the speed of the actuator. The flow rate through the orifice will also be affected by:

System pressure variations

Load pressure

Fluid density

Fluid viscosity (due to temperature changes)

If any one of these factors change during operation, the actuator speed will vary.

Therefore, fixed orifices do not provide accurate control over time and are not easily adjustable.

In order the adjust the system when changes occur, the fixed restrictions can be replaced by adjustable throttle valves as shown in Figure 2.29:

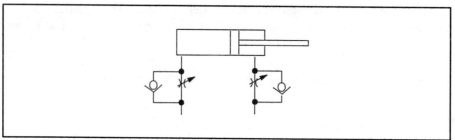

Figure 2.29

These types of valves are basically used in OPEN-loop control systems.

Because of this, actuator speed can still change during operation. The factors most likely to change the actuator's speed are:

Load Pressure Changes

System Pressure Changes

and Fluid Viscosity (Due to temperature changes)

Replacing the simple throttle valves with pressure and temperature compensated flow controls will compensate for these variations (Figure 2.30):

Pressure and Temperature Compensated Flow Control Valves

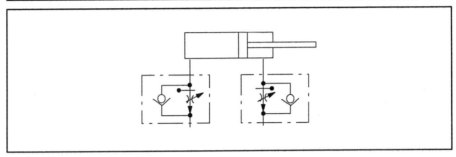

Figure 2.30

In this type of arrangement, some of the external DISTURBANCES that cause speed changes have been compensated for – others have not.

Controlling the flow rate to the actuator may not determine speed absolutely, since there may be leakage in the actuator or other system components. In other words, not all of the controlled flow may be actually driving the actuator.

Proportional and Servo Valves

While the previous valves prove to be satisfactory for a great many applications, they still require manual adjustment to give the required actuator speed.

The introduction of proportional and servo valves means that the controlled flow can now be adjusted electronically.

This allows actuator speed to be EASILY adjusted.

It also permits the acceleration and deceleration to be controlled by using RAMP GENERATORS on the valve amplifier cards (Figure 2.31):

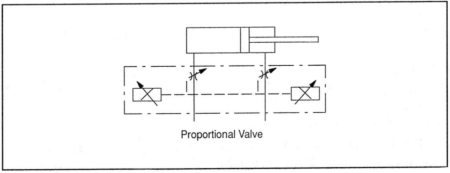

Proportional Valve

Figure 2.31

When used in an open loop system, proportional valves act as electronically adjustable restrictions.

Non-feedback, direct acting proportional valves provide a certain amount of pressure compensation, due to the way the flow forces act upon the spool. In other words, if the pressure across a non-feedback proportional valve increases, the increased pressure will attempt to push more flow through the valve. This flow increase creates a BERNOULLI force on the spool, which tends to pull the spool back and restrict the flow even more (See Vickers text on PROPORTIONAL VALVES for a more detailed explanation of the Bernoulli Force).

Feedback type proportional valves are designed to maintain their spool position for a given input signal, regardless of external changes. If a constant flow rate is required through a feedback type valve, independent of system or load pressure changes, then it must be fitted with a pressure compensating hydrostat (Figure 2.32):

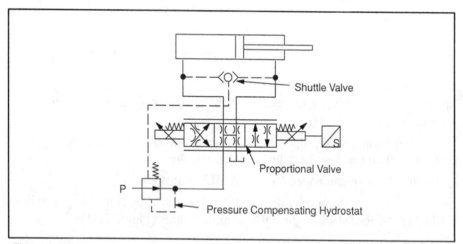

Figure 2.32

The fact that all of these valves are electrically "modulated" (adjusted) also opens up the possibility of using them in a true, closed loop velocity control system.

This can be achieved by electronically sensing the actual velocity of the actuator and feeding it back to the control amplifier in much the same way as in a position control system.

The flow rate directly affects the speed of the actuator, but in practice there is not necessarily an absolute relationship between flow rate and actuator velocity – due to leakage in the hydraulic components.

Therefore, sensing of the actuator's velocity will give a much more accurate and reliable speed control than sensing the flow rate to the actuator.

Rotary velocity can be sensed easily by using a transducer known as a tachometer-generator (or "tach-gen"). This device produces a DC output voltage that is proportional to its shaft speed.

Linear velocity can be sensed using a linear velocity transducer or by electronic conditioning of the signal from a linear POSITION transducer.

If flow sensing MUST be used (perhaps because of a difficult environment, space restrictions, etc.) then a bobbin-type flow sensor is frequently used (Figure 2.33):

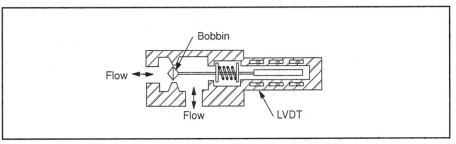

Figure 2.33

Although closed loop velocity systems are similar to closed loop position systems in concept, there are some fundamental differences between the two.

In a position control system, the amount of time that the actuator is actually moving may be relatively small. The actuator is commanded to move to a certain position and stop. While it is stopped, the valve will be centered to hold the actuator in position.

This means that the valve's NULL REGION characteristics are critical in determining how accurate the system will be, and how well it can hold its position against a reactive load.

The large deadband of a proportional valve does not give the best results in such a system, when compared to the zero-lap of a servo valve.

In a velocity control system, the control valve will spend most of its time in an operated (open) position, with flow passing through the valve. The null characteristics are less critical in this instance. Therefore, proportional valves are more suitable for closed loop velocity control than they are for closed loop position control.

The "steady-state" condition of a position control system is the condition where the load is being "held steady" in a position. The valve is centered and there is zero flow to the actuator. When the input command is equal to the feedback signal, then the error will be ZERO. This means that there is no signal to hold the valve open and create a flow to the actuator. The actuator stops at the commanded position.

The "steady-state" condition of a velocity control system is the condition where the SPEED of the load is being "held steady." The valve is OPEN, and the actuator is moving at a constant speed equal to the demand speed. Like the position control system, when the input command is equal to the feedback signal, then the error will be ZERO.

This means that there is no signal to hold the valve open and create a flow to the actuator. So if we were to use the same type of amplifier for velocity control, the actuator would stop rather than maintain its commanded speed. Therefore, a closed loop VELOCITY system requires a different type of amplifier, which is known as an INTEGRATING amplifier.

Let's compare the two types of amplifier. The amplifier normally used for a position control system is referred to as a **proportional** amplifier. The output is always directly proportional to the error (Figure 2.34):

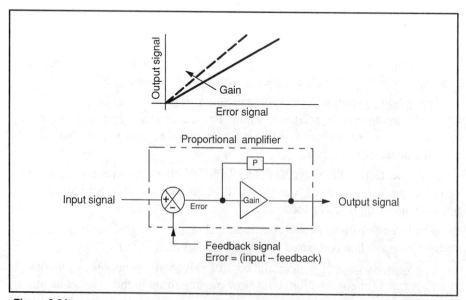

Figure 2.34

The position control system might respond to a step input command as shown in Figure 2.35:

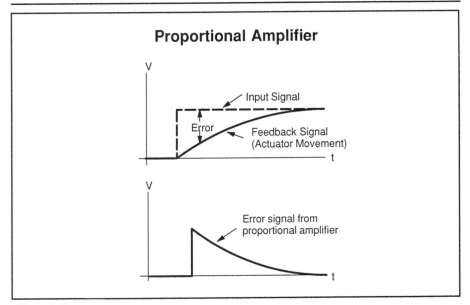

Figure 2.35

At the time just after the input signal is applied, the error signal is large, creating a large output signal to the valve.

As the actuator gets closer to commanded position, the error reduces and amplifier output reduces.

When the feedback equals the command input signal, the error will be zero and the amplifier will output zero.

An INTEGRATING amplifier produces an output proportional to the ERROR SIGNAL MULTIPLIED BY TIME (Figure 2.36):

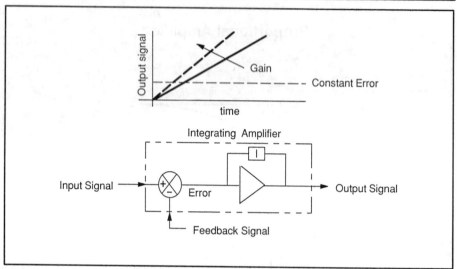

Figure 2.36

A constant command input produces a GRADUALLY INCREASING output from the amplifier (Figure 2.37):

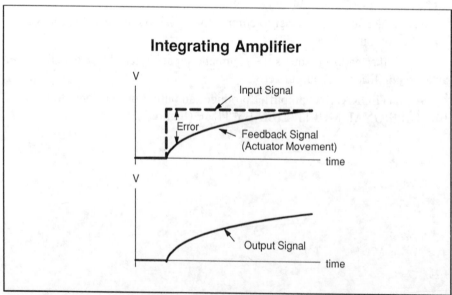

Figure 2.37

When the input signal is applied it creates a large error signal.

The output of the amplifier increases until the error is zero.

When the error is zero, the output levels off and holds constant.

In a velocity control system, when the commanded input and the actual output speeds are equal, the error signal is zero. The amplifier will then maintain a constant output level to hold the servo or proportional valve open, maintaining flow to the actuator.

Integrating amplifiers normally include a reset function, which holds the amplifier's output at zero when it is activated.

This prevents the amplifier's output from drifting. For example, if we wish to stop the actuator and keep it stopped, then we apply a zero input signal to the amplifier. But even a small amount of electrical noise on the input wire could be picked up and "integrated" by the amplifier over time, causing the actuator to start moving again.

The reset function prevents this by holding the amplifier's output at zero. The reset function is also normally used to reset the amplifier at initial system startup.

Force Control

The control of force or torque applied by an actuator is basically achieved by controlling the pressure applied to the actuator.

Open loop pressure control is typically achieved by using pressure relief or reducing valves, which provide adequate control in many applications (Figure 2.38):

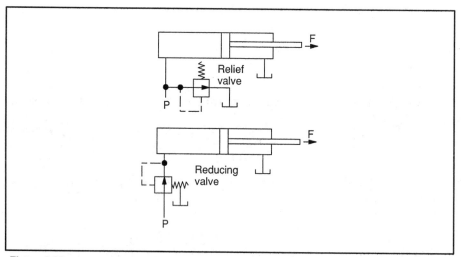

Figure 2.38

The relief or reducing valve is being used to maintain a constant pressure under varying system conditions.

In such systems, the accuracy is limited by the capabilities of the valve used. For example, variations in flow rate through the valve may affect control pressure. So might fluid viscosity changes or changes in the load applied to the system.

If it is necessary to vary the actuator force during a machine cycle, a proportional relief valve or a proportional reducing valve can be used. In these applications, the valve is normally controlled by some sort of electronic motion controller.

Such requirements are often found in press or injection molding applications, where force must be applied and removed at a controlled rate.

Greater levels of control can be achieved with proportional valves used in a closed loop system. Feedback is obtained from a pressure transducer (Figure 2.39):

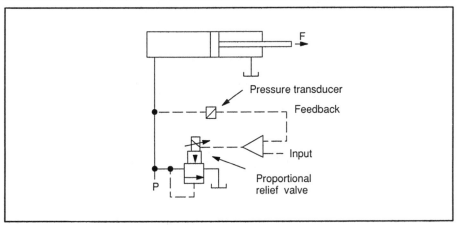

Figure 2.39

High performance proportional valves or servo valves (sliding spool type) with a zero-lapped or under-lapped spool can also be used to control actuator pressure to a high degree.

Proportional valves with overlapped spools are not ideally suited to pressure control systems because of their deadband.

However, proportional valves with overlapped spools can sometimes be used successfully in the same manner that you would use a proportional relief valve (Figure 2.40):

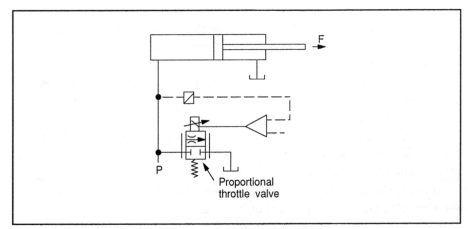

Figure 2.40

A typical arrangement for a closed loop force control system is shown in Figure 2.41:

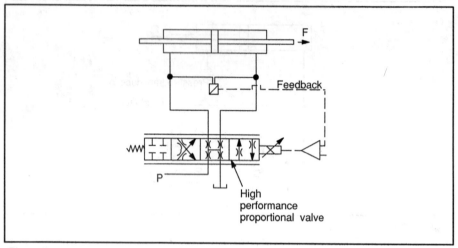

Figure 2.41

The command input signal is summed with the feedback signal to produce the error signal.

To create a feedback signal, the output of the system can either be sensed directly with a load cell or force transducer, or it can be sensed indirectly by measuring the pressure applied to the actuator.

In many applications, it may be more convenient to sense pressure instead of force. However, if pressure sensing is used, the effects of back-pressure on the actuator outlet have to be taken into account, since this will tend to subtract from the total output force.

If an equal area actuator is used (a motor or a double rod cylinder, for example), a differential pressure transducer will automatically compensate for back pressure, as shown in Figure 2.42.

If a differential area cylinder is used, two pressure transducers will be needed, with one transducer's output suitably scaled to account for the area difference between the two sides of the piston.

Leakage in the control valve or actuator will tend to cause a steady-state error. This means that the control valve will have to be partially open in order compensate for this leakage.

In order for the valve to be partially open, it must get a drive signal from the amplifier, meaning that an error signal must exist.

This is also true if the actuator is moving at the same time that the pressure is being controlled. The spool has to be shifted over a certain amount in order to allow flow to the actuator – once again requiring an error signal to exist.

Increasing the gain will tend to reduce this error, but will not eliminate it. This is a situation that is similar to the closed loop velocity system discussed earlier. An integrating amplifier was used to build up the output signal and level off when the error reduced to zero.

Such an arrangement could also be used in a pressure control system, but it is normally combined with a proportional amplifier to form what is called a PI type amplifier (proportional plus integral).

The proportional part provides quick response to the input command, and the integral part eliminates the steady-state error (Figure 2.42):

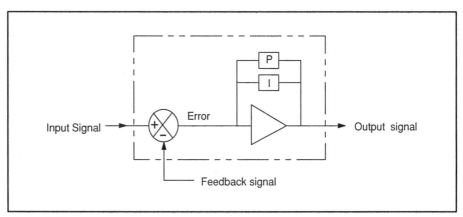

Figure 2.42

Another advantage of using a spool type valve in this way is that the actuator frequently needs to be both position-controlled and force-controlled.

For example, in a press application, the initial pre-form motion may need to be position-controlled and the actual pressing operation may require force control. The same control valve can be used for both operations by switching from a proportional to an integral amplifier at the required time during the machine cycle (Figure 2.43):

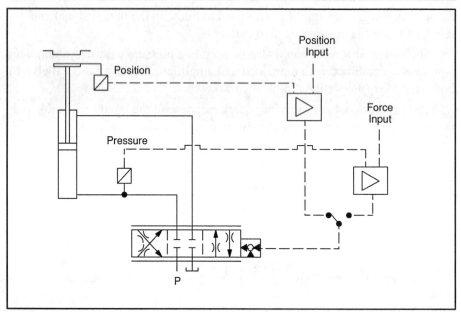

Figure 2.43

Closed Loop
Control Components

C H A P T E R

3

Many of the components used in closed loop hydraulic systems will be the same as those used in open loop applications (pumps, relief valves, filters, etc.) and should be well understood.

This chapter will be confined to those specialized components that are normally found only in closed loop systems.

Control Valves

Closed loop systems have traditionally used servo valves as the hydraulic control valve. Originally developed for aerospace applications, servo valves have been adapted for industrial use over the years, and are now used extensively in industry.

Recently, proportional valves have become available with a performance approaching that of servo valves. Due to their lower cost, they are becoming increasingly popular.

The distinction between a proportional valve and a servo valve is becoming more and more difficult to define. The only major difference between them is the method of spool actuation. Proportional valves are operated by a solenoid which requires a relatively large power input, as opposed to the force or torque motor in the servo valve, which requires little power input.

Both servo valves and spool-type proportional valves are basically direction-and-flow control valves with four main working ports (Figure 3.1):

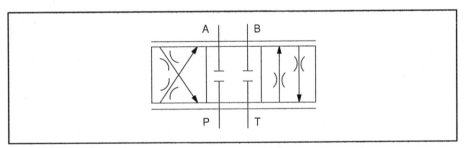

Figure 3.1

The direction of spool motion determines direction of flow, and the amount of spool motion determines flow rate through the valve.

A typical proportional valve for closed loop applications is shown in Figure 3.2:

Closed Loop Proportional Valves

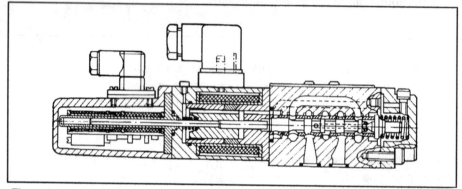

Figure 3.2

The valve uses a standard directional valve body with a spring-biased sliding spool inside. In order to achieve accurate matching of the spool lands to the valve ports, the body is fitted with a sleeve. The ports in this sleeve are accurately machined to match the spool lands. The spool then slides inside of this sleeve.

The spool is positioned by means of a single proportional solenoid. Also connected to the solenoid is the LVDT, which provides a spool position feedback signal to the valve's amplifier. This allows the loop to be closed on the spool itself, providing very accurate control of spool position.

The spool is "biased" with a spring to push the spool fully over, providing a fail-safe cutoff condition when no power is applied.

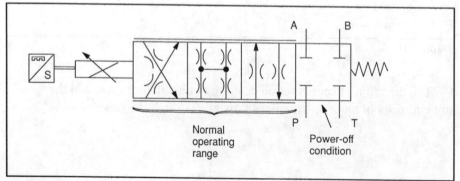

Figure 3.3

Servo Valves

Although several different types of servo valves are used in industrial applications, the most common is the **flapper/nozzle** type shown in Figure 3.4:

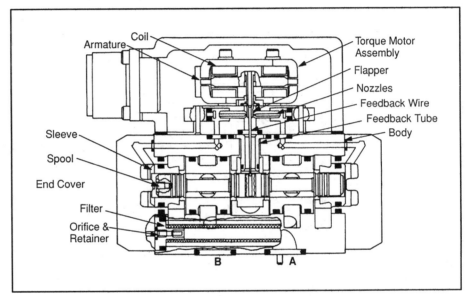

Figure 3.4

This valve is a two stage device. In other words, it is pilot operated. The pilot stage flapper is operated by a torque motor.

The flapper/nozzle assembly controls the pressure on either end of the main spool, causing it to move. Movement of the main spool, in addition to regulating flow through the valve, also creates a feedback to the pilot stage in the form of a mechanical movement of the flapper.

The operation of this valve is more completely described in the following pages.

Torque Motor

The torque motor consists of two permanent magnets, each surrounded by a frame (Figure 3.5). The frames and magnets are connected in such a way that the upper frame is polarized NORTH and the lower frame is SOUTH.

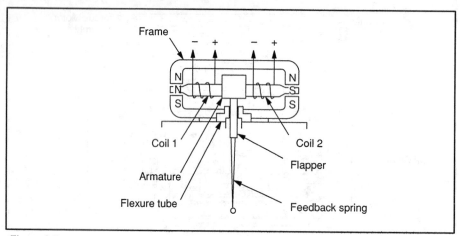

Figure 3.5

The armature/flapper assembly is supported between the two frames by a FLEXURE TUBE. This tube allows the armature to move and also seals hydraulic fluid out of the electrical portion of the valve.

Surrounding the armature on both sides are two coils, one on each side of the flexure tube. Passing an electrical current through the coils causes the armature to be magnetized as shown in Figure 3.6:

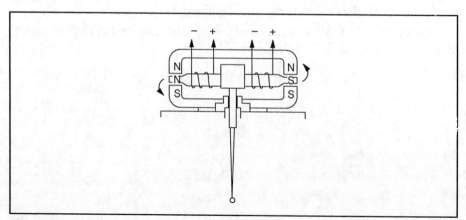

Figure 3.6

The magnetic poles formed in the armature will be attracted to the poles in frame as shown. This will cause the entire armature to twist, and move the flapper.

With coil current as shown, the armature would twist counterclockwise and move the flapper to the right.

Reversing the coil current would reverse the magnetic field in the armature, causing the flapper to move to the left.

Nozzle and Flapper Assembly

Attached to the center of the armature is a flapper and feedback spring that extend down through the flexure tube.

A ball on the end of the feedback spring fits into a recess in the spool.

On each side of the flapper are two matched nozzles that are supplied with pressurized fluid (Figure 3.7):

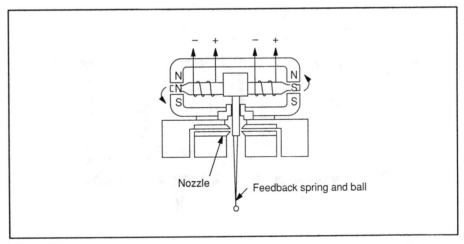

Figure 3.7

The pressure feed for each nozzle comes from the supply pressure port on the main stage, by way of a built-in filter and fixed orifices.

The fixed orifices limit the amount of flow to the nozzles.

The filter is a "last-defense" against fluid contamination particles attempting to enter this sensitive section of the valve.

In some cases, it is advantageous to supply the pilot pressure to the two nozzles with a separate pilot pressure supply which is connected to the valve's "fifth port." (For example, when the system pressure is too low for good servo valve response).

The pressure at each end of the main spool is now affected by the amount of restriction of the nozzles; and restriction of the nozzles is controlled by the position of the flapper.

Here's how it works:

With zero electrical signal to the coils in the torque motor, the flapper will be centered between the two nozzles. This creates equal restriction of the nozzles and equal pressures on the ends of the spool.

The main spool will be centered at this time. If it were not centered, the flapper would be moved by the feedback spring and unbalance the spool end pressures.

When an electrical signal is applied to the torque motor, the armature twists, causing the flapper to block one of the nozzles.

In Figure 3.8, we show the flapper blocking off the right-hand nozzle.

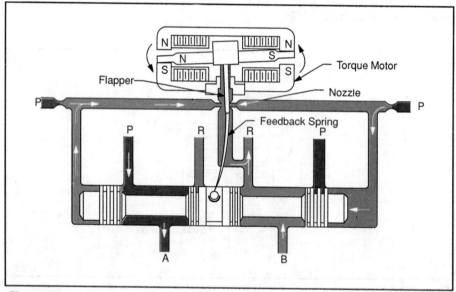

Figure 3.8

This creates a greater restriction in the right-hand nozzle and less restriction in the left-hand nozzle, and unbalances the pressure on the ends of the spool. Pressure will increase on the right side, and decrease on the left side, causing the spool to be pushed to the LEFT.

The main stage now ports P to A and B to Return (Tank). But that's not all there is to it!

As the main spool moves to the left, it pulls the feedback spring with it. The feedback spring pulls the flapper away from the blocked nozzle, drawing the flapper back toward its centered position.

When the flapper is again centered, the pressure on the ends of the spool equalize again (Figure 3.9):

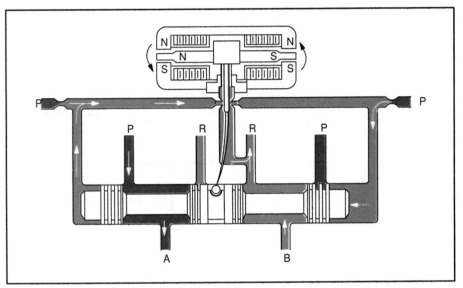

Figure 3.9

The main spool continues to move to the left until the force placed on the flapper by the torque motor EQUALS the force placed on the flapper by the feedback spring. When these two forces balance, spool motion stops.

Since the force created by the torque motor is proportional to coil current, the amount of spool movement is also determined by coil current.

We can move the spool to the right by reversing the direction of the electrical current sent to the coils.

Main Stage

In order to achieve accurate matching of the spool lands to the valve ports, the body of the valve is fitted with a sleeve that has spark-eroded flow ports.

The spool slides into this sleeve, which provides very accurate flow openings relative to input current.

There is also a null adjustment assembly fitted in the body. By manually setting this adjuster, the sleeve can be displaced relative to the spool, allowing the sleeve to be precisely centered on the spool lands. It is typically adjusted with zero input current applied to the torque motor (Figure 3.10):

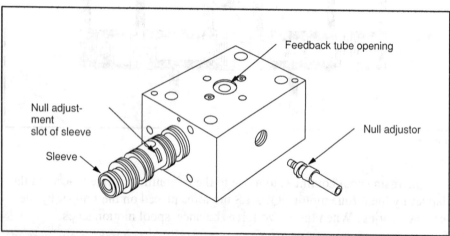

Figure 3.10

Lap Condition

One vital feature of a control valve is the relative position of the edge of the spool land and the edge of the port opening, when the valve is in the null (centered) position.

This is referred to as the **SPOOL LAP CONDITION**.

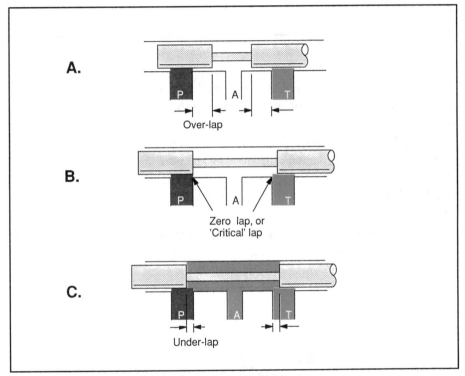

Figure 3.11

Figure 3.11-A shows an OVERLAPPED spool. The spool land overlaps the edge of the port.

Figure 3.11-B shows a ZERO LAPPED spool, where the spool lands align exactly with the edge of the ports.

Figure 3.11-C shows an UNDERLAPPED spool. The edge of the spool does not fully cover the port in the null position.

Amount of overlap or underlap is normally expressed as a percentage of total spool movement, or as a percentage of full input signal.

Let's take a closer look at the OVERLAPPED spool in Figure 3.11-A.

Because of the overlap, the spool must be moved a certain amount from null before flow through the valve can even begin to occur. This effect is called the DEADBAND of the valve (Figure 3.12):

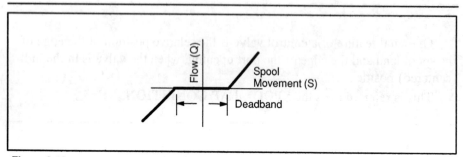

Figure 3.12

Once the spool has cleared the port, the flow rate (Q) will be directly proportional to the amount of spool movement (S) for a given pressure drop across the valve.

The relationship between spool movement and flow may not always be a linear one. This will depend on the shape of the ports in the sleeve and whether or not notches are used in the spool lands. Generally, square ports and an unnotched spool will give a direct relationship between flow and spool movement.

Since spool movement is directly related to input signal, we can also chart the graph in Figure 3.13, relating FLOW to input signal.

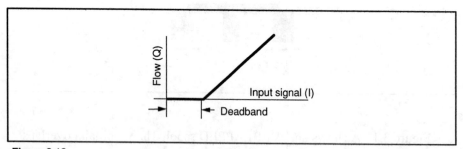

Figure 3.13

In a valve with an overlapped spool, there will be a minimum input signal required in order to create a flow through the valve. This means that any signal less than this minimum level will not open the valve, and the system will not be very sensitive to small input or error signals.

Flow Gain

Once the input signal is large enough to create a flow through the valve, the flow rate will be proportional to the input signal, as shown in Figure 3.14:

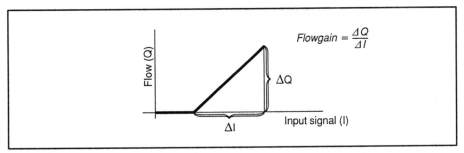

Figure 3.14

The steepness of the slope charted above is referred to as the FLOW GAIN of the valve. This is simply the change in flow rate over a given change in input signal, which is measured at a specific pressure drop across the valve. Note the presence of deadband in Figure 3.14, indicating that this is a graph for an overlapped spool.

With Zero Lapped spools (Figure 3.15) the edge of the spool land lines up precisely with the edge of the port, eliminating any deadband.

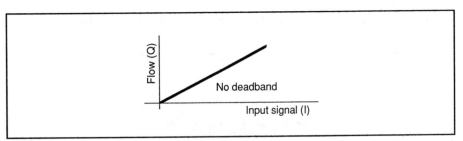

Figure 3.15

Flow through this valve will occur even at very low signal levels, making the valve very responsive around the null position.

The flow gain may vary from its nominal value around the null position due to manufacturing tolerances and the squareness of the ports (Figure 3.16):

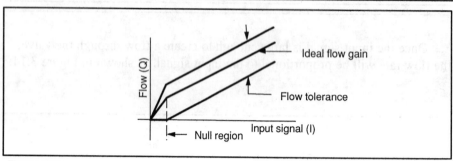

Figure 3.16

For the underlapped spool, the flow vs. signal graph for one of the ports would look like Figure 3.17:

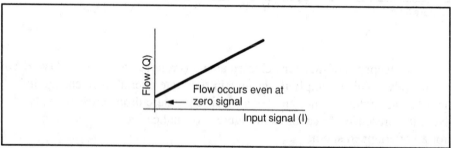

Figure 3.17

There would be a small amount of flow though the port, even at zero input signal, since the spool does not fully close off the port at null position.

However, in actuality, both ports have to be considered (Figure 3.18):

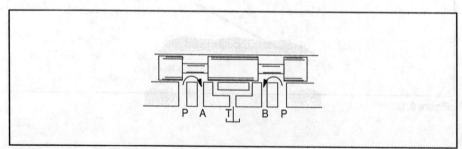

Figure 3.18

If we assume an equal underlap on both sides, both ports will be partially open in the null position.

This will equalize the pressure to our actuator in both A and B ports, and will prevent flow as a result.

As the spool is moved across, one flow path opens up and the other closes off. The flow gain will therefore be as shown in Figure 3.19:

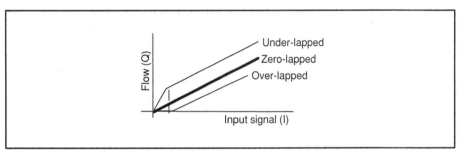

Figure 3.19

The result is that an underlapped spool will have an unusually high flow gain around the null region.

This characteristic may be highly desirable in some applications.

Pressure Gain

Another important feature, determined by the lap condition, is known as the **PRESSURE GAIN** of the valve.

It is defined as the rate of change of output pressure with input current, and it is measured with zero flow (A & B ports blocked).

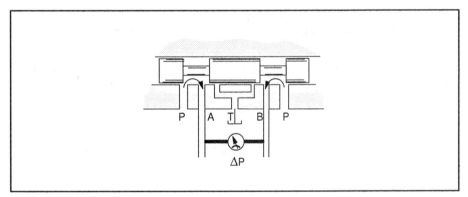

Figure 3.20

With the spool in null position, as shown in Figure 3.20, there will be some leakage from both pressure (P) ports to the tank (T) ports.

This means that the pressure in the A and B ports will be at some pressure between supply and tank line pressure.

Assuming an ideal valve, the pressure in the A and B ports will be equal, so the differential pressure gauge will read 0 PSI.

If the spool is now moved toward the left, the A port will be opened to P and the B port will be opened to T. The A port pressure will rise and the B port pressure will fall until Port A reaches supply pressure and Port B drops to 0 (or whatever tank line pressure actually is).

At this point, shown in Figure 3.21, the differential pressure gauge will read full supply pressure (assuming tank line pressure = 0).

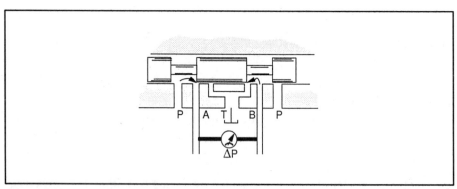

Figure 3.21

For a zero lapped spool, full differential pressure will be reached when the input signal (i.e. spool movement) reaches only 3% to 4% of maximum.

Pressure gain is the slope of this pressure increase, and can be specified either graphically or numerically, (for example, 30% of supply pressure per 1% of rated current).

Figure 3.22 shows representative pressure gain graphs for the various lap conditions we have discussed.

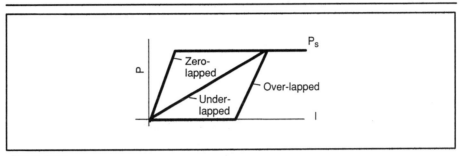

Figure 3.22

With an underlapped spool, a greater amount of spool movement is required before the tank port closes off fully. This results in a shallower slope and a lower value of pressure gain.

In an overlapped spool, pressure will not rise until the spool has passed the deadband region.

The pressure gain of a valve is obviously a very important criteria for a closed loop pressure control application. It is also important in determining what is called the steady-state accuracy of a position control system. We will examine this later, in more detail.

However, as you can see in Figure 3.22, the zero lapped spool will give the best results in applications requiring a high pressure gain with minimal deadband.

Hysteresis

As in any electromechanical device, friction and magnetic effects cause a condition called HYSTERESIS.

The output flow of the valve will differ, depending on whether the input signal is increasing or decreasing (Figure 3.23):

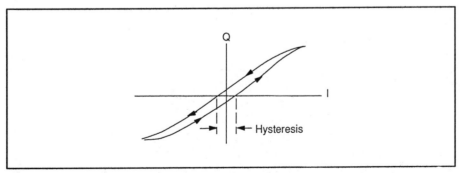

Figure 3.23

Hysteresis is expressed as a percentage of maximum rated input signal.

Threshold

Another characteristic related to hysteresis is called THRESHOLD, or reversal error.

This is defined as the difference in input current that is required to change from an INCREASING FLOW to a DECREASING FLOW, or vice versa (Figure 3.24):

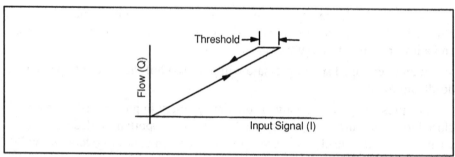

Figure 3.24

Like hysteresis, threshold is normally quoted as a percentage of maximum rated input signal.

Linearity and Symmetry

Two other characteristics that relate the flow vs. input signal behavior of a servo valve to the ideal flow gain condition are **LINEARITY** error and **SYMMETRY** error.

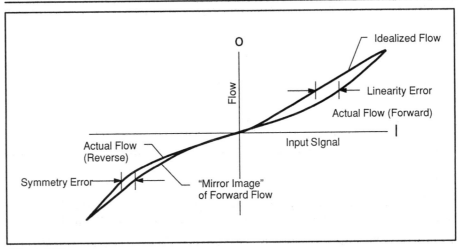

Figure 3.25

Linearity error is shown above. First, the point is found where maximum difference occurs between actual flow rate and the ideal flow gain line. The gap between the ideal flow and actual flow yields a certain amount of linearity error that is expressed as a percentage of maximum rated current.

$$Linearity\ error \quad = \quad \frac{gap\ width\ (in\ milliamps)}{max.\ rated\ input\ signal\ (in\ milliamps)} \times 100\%$$

Symmetry error is the difference between the flow gain lines for spool movement on either side of center. This is expressed as a difference in <u>flow gain</u> between the two sides as a percentage of whichever flow gain is greater.

Flow Rating

A servo or high performance proportional valve is basically a combined direction and flow control valve. Flow rate is controlled by adjusting the spool/port opening, which creates a variable sized hole, or ORIFICE, through which the hydraulic fluid passes.

Since there are two flow paths through the valve, in most applications the servo valve will create two restrictions through which fluid must flow (Figure 3.26):

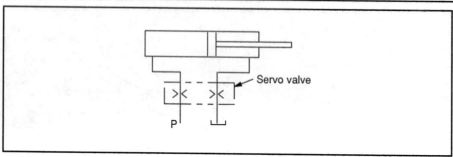

Servo valve

P

Figure 3.26

It has been found, through laboratory experiments, that the variable restrictions in a servo valve behave like a SHARP EDGED ORIFICE.

This is convenient because, in sharp edged orifices, a neat mathematical relationship exists between the flow rate through the orifice, the size of the orifice and the pressure drop across the orifice:

$$ Q \ :: \ A \sqrt{\Delta P} $$

Equation #1

The equation says that flow through a sharp edged orifice is
    Proportional to the area of the orifice, and
    Proportional to the <u>square root</u> of the pressure drop across the orifice.

In other words, if you increase the area of the orifice (by opening the valve further), the flow through will increase by a directly proportional amount.

If you can also increase the flow through the orifice by increasing the pressure drop across it. However, it does not increase by a directly proportional amount, due to the square root relationship.

For example, if you wish to DOUBLE the flow through the orifice while leaving the area the same, then you must QUADRUPLE the pressure drop across it.

Spool movement, and the resulting change in orifice area, is controlled by the input signal to the servo valve's coils.

The pressure drop across the valve is determined by:
    The system pressure (P_S) that is available at the valve,
    Minus the pressure required to move the load (P_L),
    Minus the return line pressure (P_T).

In equation form, $\Delta P_V \ = \ P_S - P_L - P_T$

Traditionally, the rated flow of a servo valve is quoted at 100% input current and a 1000 PSI total pressure drop across the valve.

Each flow path (P to A and B to T) can be considered to drop 500 PSI each, when operating under RATED conditions.

However, we don't always have the luxury of a system that produces exactly 1000 PSI of pressure drop across the valve. We may get more or less than the rated flow through a given servo valve depending on the system conditions under which the valve is operating.

We can accurately predict the flow through a servo valve at a pressure drop other than 1000 PSI by using the following DE-RATING EQUATION:

$$Q_A = Q_R \ \frac{I_A}{100\%} \ \sqrt{\frac{\Delta P_V}{1000}}$$

Equation #2

Where: Q_A = ACTUAL flow rate through valve (in GPM)
Q_R = RATED flow of the servo valve (in GPM)
I_A = ACTUAL input signal level (as a percentage of full input signal level)
ΔP_V = Total Pressure Drop across valve (in PSI)

When we chart the performance of a servo valve in the lab, we find that it follows this mathematical relationship very closely, as graphed in Figure 3.27:

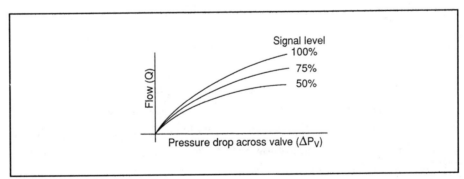

Figure 3.27

The flow charted in the graph is the ACTUAL flow through the valve, at various input signal levels.

If we chart the relationship between flow and pressure drop on a log-log scale, we find that it becomes a straight line graph as shown in Figure 3.28:

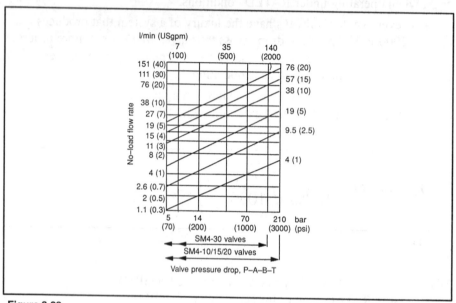

Figure 3.28

This graph charts flow vs. pressure drop for 6 different sizes of servo valves. Note that the flow through the valve equals the flow rating of the valve only at 70 Bar (1000 PSI) pressure drop.

Example 3.1:

For a servo valve rated at 10 GPM, how much flow will we get through the valve at 500 PSI drop and 75% input signal?

Look at Figure 3.28 and find the graph line for the 38 l/min (10 GPM) valve. The graph lines are labeled by size on the right-hand side of the graph.

Follow the graph-line for this valve down to where it intersects the vertical 35 bar (500 PSI) pressure line (this line is labeled at the top of the graph).

At the intersection point, follow the horizontal flow line to the left side of the graph and read 27 l/min (7 GPM).

This means that our 10 GPM will only pass 7 GPM at 500 PSI drop and 100% input signal. To find the flow at 75% input signal, multiply 7 GPM x.75 to arrive at the actual flow of 5.25 GPM.

We can also use the equation to directly calculate the actual flow.

$$Q_A = 10\,GPM \times \frac{75\%}{100\%} \times \sqrt{\frac{500\,PSI}{1000\,PSI}}$$

$$Q_A = 10 \times .75 \times \sqrt{.5} = 7.5 \times .7071 = 5.3\,GPM$$

The equation will usually give a more accurate flow value than the graph.

Historically, servo valves have been operated at a higher pressure drop than conventional solenoid valves or proportional valves. This is because a higher pressure drop enables the use of a smaller valve (size was a vital consideration in aerospace applications); and generally, the smaller the valve, the faster its response.

Response time of a servo valve can also be improved by increasing system pressure. In applications where response time is critical, a relatively high pressure drop will improve the system stiffness and accuracy.

In applications where efficiency is more important than valve response, a lower pressure drop can be used. However, the minimum pressure drop at which a flapper/nozzle type servo valve will operate is typically 200 PSI.

In systems where the system pressure is too low to directly operate the pilot stage of a servo valve, an alternative is to supply a high pilot pressure to the valve separately, via another small pump, to the fifth (pilot) port of the servo valve.

You should also be aware that there is a specific relationship between valve pressure drop and MAXIMUM POWER TRANSMISSION through a servo valve. Hydraulic POWER is a function of FLOW and PRESSURE.

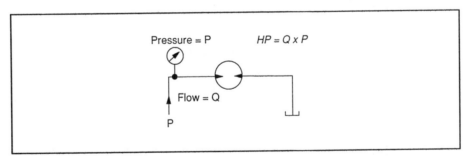

Figure 3.29

Normally, all of the flow through the valve goes to or from the actuator, so actuator flow is the same as valve flow.

As we said earlier, this flow is proportional to the square root of the pressure drop at a given input signal.

Increasing the pressure drop will create a higher flow rate to the actuator. This will tend to INCREASE available power to the actuator.

But at the same time, the higher pressure drop will DECREASE the pressure available to drive the actuator, and will therefore tend to REDUCE available power to the actuator.

Initially, increasing the pressure drop will INCREASE the power available to the actuator. The higher flow will outweigh the power lost by the increased drop across the valve.

However, after a certain point, the power lost due to the pressure drop becomes larger than the power gained by higher flow to the actuator! In fact, this "point of diminishing returns" can be mathematically found (Figure 3.30):

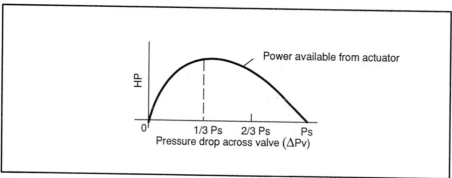

Figure 3.30

For a given size of valve, maximum power will be transmitted to the actuator when the valve pressure drop is equal to ONE-THIRD of the system pressure.

FOR MAXIMUM POWER TRANSMISSION

$$\Delta P_V \; = \; \frac{1}{3} P_S$$

Dynamic Characteristics

In both Servo Valves and feedback type Proportional Valves, the valve spool itself is positioned by means of a closed loop control mechanism (sometimes referred to as an "inner loop").

For a feedback type proportional valve, the input signal is fed to the valve's amplifier, which produces a drive current to the valve solenoid. A feedback signal from the spool's position sensor (LVDT) is summed with the input signal.

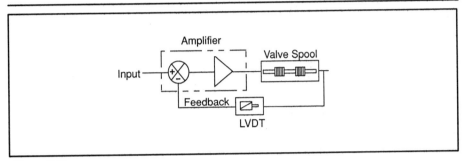

Figure 3.31

Any error in the actual position of the spool caused by friction or flow forces is automatically corrected.

In a flapper/nozzle servo valve, the mechanism is quite different. The feedback signal is provided mechanically by the action of the spool pushing on the feedback spring/flapper assembly. In this case, the flapper acts as the "summing junction," with the input "signal" coming from the torque motor armature and the feedback coming from the main spool motion.

If you consider the valve itself as a closed loop system, the dynamic characteristics can be described by examining its response to either a STEP input or a SINUSOIDALLY VARYING (sine wave) input.

Step Response

Step response of a valve is, simply, the response of a valve to a step change input. A step input is an input signal that is suddenly "switched on" as shown in Figure 3.32:

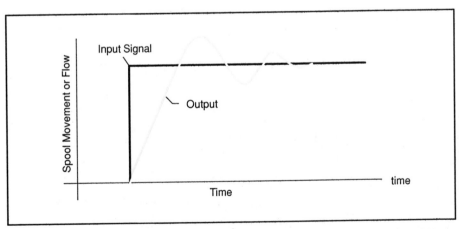

Figure 3.32

The manner in which the output (spool movement) responds in the above figure is typical for a closed loop position control system. The spool moves rapidly toward the commanded position, overshoots a certain amount, then gradually settles at the commanded position after a series of decreasing oscillations.

The actual time that this takes is usually a matter of a few milliseconds.

It is therefore not altogether straightforward to define what is meant by the "step response" of a valve. Some of the methods used to define step response are:

1. The time it takes to reach 100% of demand position, ignoring the fact that the spool will overshoot.
2. The time to travel from 10% to 90% of demand position, ignoring initial and final irregularities.
3. The time to reach 63% (one time-constant) of commanded position.

In some cases, a "settling time" is also quoted, which is the amount of time it takes for the oscillations to die down within 5% of the commanded position.

The step response of a servo valve is usually an unnecessary academic concern, since the servo valve's response is usually much faster than the system it is controlling.

However, in fast-response pressure control applications, settling time can become a very important factor.

Frequency Response

A more useful measure of valve response is to look at its **FREQUENCY RESPONSE** to a sine wave input signal.

When the input signal is varying at a very low frequency, the spool of the valve will be able to follow the input signal very closely (Figure 3.33):

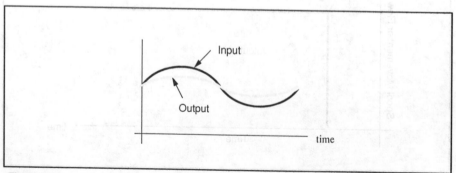

Figure 3.33

However, if we speed up the frequency of the input signal, the spool will be less and less able to move as rapidly as the input.

The output (spool movement) starts to lag behind the input, and the output is not able to reach maximum value before the input reverses (Figure 3.34):

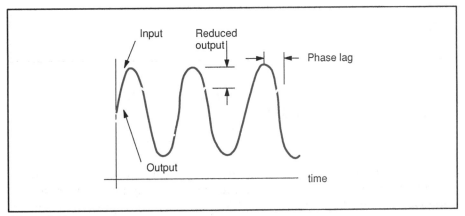

Figure 3.34

This lag between output and input is referred to as **PHASE LAG** or **PHASE SHIFT**.

The reduced level of output which occurs at higher frequencies is known as **ATTENUATION**.

Phase lag and attenuation are normally presented in servo valve specification sheets by means of a graph.

Attenuation is measured in a unit known as the DECIBEL (or dB):

$$dB = 20 \log\left(\frac{Output}{Input}\right)$$

Equation #3

In control valves, Input is measured in terms of current (milliamps) but Output is measured in terms of flow rate (GPM) or spool movement (inches).

To avoid confusion, and make the units the same in all cases, input and output are measured in terms of a percentage of their maximum values:

Input = Percentage of maximum input signal
Output = Percentage of maximum output signal

When the input signal is varied at a very low frequency, the input and output magnitudes will be the same, since the spool is easily able to follow the input signal.

$$\frac{Output}{Input} \;=\; 1$$

Equation #4

The attenuation will therefore be:

$$dB \;=\; 20 \log 1 \;=\; 20 \times 0 \;=\; 0$$

As the input signal increases in frequency, the output becomes less and less able to follow the input. We can represent this on a graph known as a **Bode Diagram** or **Bode Plot** (Figure 3.35), where attenuation vs. frequency is plotted.

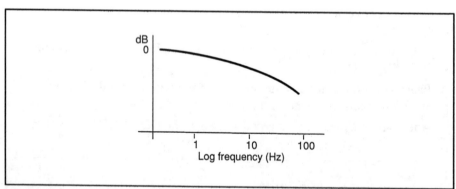

Figure 3.35

As you can see, at low frequencies the attenuation is close to 0 dB (no attenuation). But at higher frequencies, the curve starts to slope downward, indicating more and more attenuation of the output.

One important point on the attenuation curve is the –3 dB point. This is the point where the output has dropped to 70% of its maximum, as shown below:

$$dB \;=\; 20 \log\!\left(\frac{70\%}{100\%}\right) \;=\; 20 \log(.7)$$

$$dB \;=\; 20 \times (-0.15) \;=\; -3$$

The minus sign indicates an attenuation loss, rather than a gain. This point is normally referred to as being "3 dB down."

The input frequency that creates an attenuation of 3 dB is one defining characteristic of a valve and is referred to as its **BANDWIDTH** (Figure 3.36):

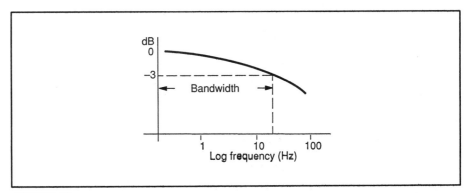

Figure 3.36

The bandwidth of a component is that frequency at which the output is attenuated to about 70% of maximum, or -3 dB.

As the input frequency increases, the output will also tend to exhibit phase lag, as we said earlier. This can also be graphed on a Bode Plot as shown in Figure 3.37:

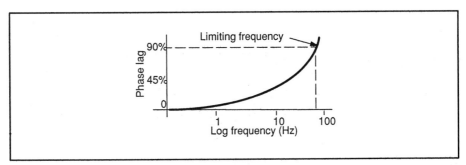

Figure 3.37

At low frequencies, the phase shift is very small, but it increases as the input frequency increases.

The frequency where a component exhibits a 90 degree phase lag is another standard characteristic, referred to as its **LIMITING FREQUENCY**.

The limiting frequency is the normal characteristic that is used to compare the performance of control valves, or for selecting a suitable valve for an application, because the limiting frequency is affected by both supply pressure and input signal amplitude.

In any case, it is important, when comparing the performance of valves, to ensure that the parameters being compared are the same, and that they are derived in the same manner.

Amplifiers

The purpose of the amplifier, or controller, in a closed loop system is basically to sum the command input and feedback signals to produce an appropriate drive signal for the valve.

The amplifier module consists of a summing junction and a signal amplifier as shown in Figure 3.38:

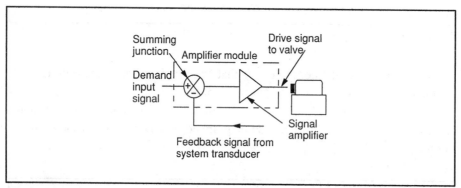

Figure 3.38

Figure 3.39 shows a simplified schematic of an amplifier module, with its important adjustments shown.

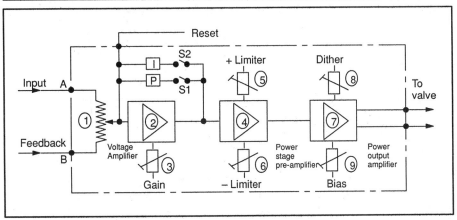

Figure 3.39

The input command signal and the transducer feedback signal are connected to terminals A and B across potentiometer (1) which acts as the summing junction. The input and feedback voltages must be of opposite polarity.

The potentiometer, when used in this manner at the input, is referred to as a RATIO pot because it allows you to establish a proper ratio between command and feedback voltages of differing ranges.

For example, if the command input was +10V and the feedback was –10V, the ratio pot would be adjusted to the center position. If the command input was +10V but the feedback signal was –20V, the pot would be adjusted down 1/3 from the top, to balance the effects of the two signals and produce a zero volt signal at the input to the amplifier.

The wiper voltage would be proportional to the sum of the two signals, and would become the needed ERROR SIGNAL.

Voltage amplifier (2) can operate in either linear mode by closing switch S1, or in integrating mode by closing S2. Linear mode is used for position control applications, whereas integrating mode is used for velocity control systems. In either situation, potentiometer (3) adjusts the gain of the voltage amplifier within its selected range.

In linear mode, the gain determines the AMOUNT OF OUTPUT VOLTAGE for a given input voltage (Figure 3.40a):

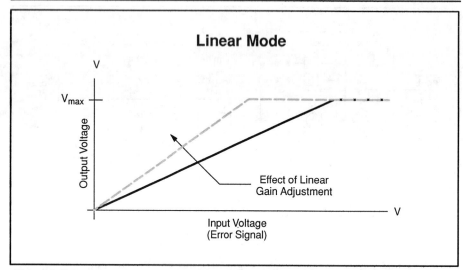

Figure 3.40a

In integrating mode, the gain determines the RATE OF OUTPUT VOLTAGE INCREASE IN VOLTS/SEC for a given input voltage (Figure 3.40b):

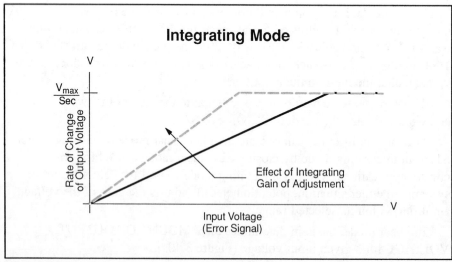

Figure 3.40b

When integrating mode is used, a reset function is provided at terminal C. When a ground is applied at C, the reset function holds the output of the amplifier at ZERO. This is normally done when the amplifier's input is at zero, to prevent electrical noise from causing the amplifier's output to drift and therefore to prevent unwanted actuator movement.

The reset function is also used during startup, to prevent actuator movement while voltages are stabilizing during initial power-on.

The output of the voltage amplifier drives another amp called the power stage pre-amplifier (4). This stage has adjustments called limiters (+ and −) which limit the maximum drive current which the module will supply to the valve in either direction. This adjustment is made to prevent burning out the valve coils through amplification of an unusually large input signal.

The final stage of the amplifier module is the power output stage (7), which creates the actual drive current to the servo valve's torque motor coils.

This amplifier uses a CURRENT FEEDBACK arrangement to sense the amount of CURRENT transmitted to the valve. Current feedback is used to overcome the fact that the resistance of the coils will change with temperature. The power stage will produce a certain amount of current for a given voltage input, regardless of coil resistance and inductance variations.

Current feedback also improves the response of the system by minimizing the effects of coil inductance on the output signal.

This stage has a DITHER adjustment (8) to reduce hysteresis and prevent silting of the valve spool. It is normally adjusted as high as possible without causing noticeable motion of, or noise in, the actuator.

This stage also has a BIAS adjustment (Figure 3.41) which can be used to offset the output, if desired.

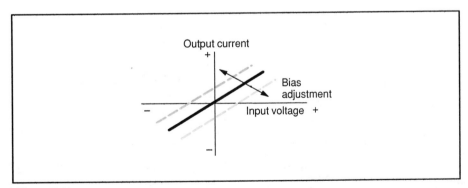

Figure 3.41

This adjustment is useful in those applications where a positive or negative output is needed while the input is ZERO. It also tends to be used as an electronic valve null adjustment.

Ramp Modules

The addition of a ramp module to a valve amplifier enables the system output to move from one condition to another at a pre-set rate.

In a position control system, for example, the speed of movement from one position to another can be controlled by the ramp module.

In a velocity control system, acceleration and deceleration can be controlled by a ramp module.

In a pressure control system, the rate of pressure rise and fall can be controlled (Figure 3.42):

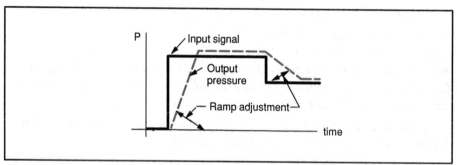

Figure 3.42

A simplified schematic of a ramp module is shown in Figure 3.43, which incorporates two ramp angle potentiometers.

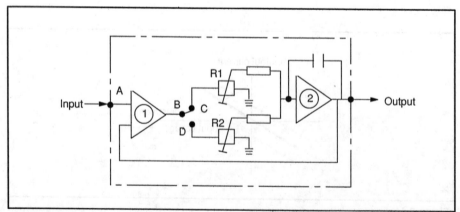

Figure 3.43

A step change in input signal at connection A causes high-gain amp (1) to go to maximum output. Assuming connection B is switched to connection C as shown, the output of the high-gain amp is fed to ramp potentiometer R1.

R1 provides an adjustable voltage which is fed to the integrating amplifier (2). The amount of input voltage (set by R1) determines the ramp angle.

The output of the ramp generator is an input to the valve's amplifier, and is also fed back to the high gain amplifier (1).

When the ramp generator's output reaches the same voltage level as the input step voltage, the high gain amp (1) produces zero output, causing the ramp generator's output to level off and hold (Figure 3.44):

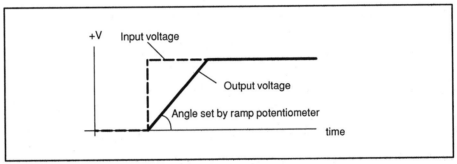

Figure 3.44

The ramp module shown in Figure 3.43 can provide two different ramp rates, depending on the position of the switch and the adjustment of R1 and R2.

To create multiple ramps, additional pots and/or capacitors can be connected to the module and activated by switches or relays. Ramp modules can also be combined with selectable command input pots as shown in Figure 3.45:

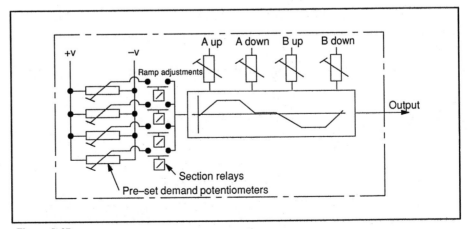

Figure 3.45

Here we see a ramp module with four different command input pots, each with its wiper connected through a relay contact. By energizing the appropriate relay, we can supply any one of four different input voltages to the ramp module, giving us four possible ramp rates from the module.

Transducers

One identifying feature of a closed loop system is the fact that the state of the controlled variable (whether it is position, velocity or force) is fed back to the control amplifier for **AUTOMATIC ERROR CORRECTION**.

This requires a device that will measure the position, velocity, pressure, torque or force, and convert it into an electrical signal that the control amplifier can use.

Such a device is called a **FEEDBACK TRANSDUCER**.

There are TWO stages in obtaining a feedback signal:

First, the transducer has to be able to sense the quantity that we are measuring.

Second, the transducer signal has to be converted into a signal that our control amplifier can use.

This process of signal conversion is called SIGNAL CONDITIONING.

Signal conditioning may include:

Amplification – increasing the power of the signal
Buffering – keeping it isolated from the amplifier
Demodulation – converting an AC signal to DC
Conversion from Voltage to Current – or vice versa
Conversion from Digital to Analog – or vice versa
Calibration – scaling the feedback to match the command

Some transducers have electronic integrated circuitry built in, which performs the necessary conditioning. Others require separate conditioning devices for use.

Consider a simple block diagram of a closed loop system (Figure 3.46):

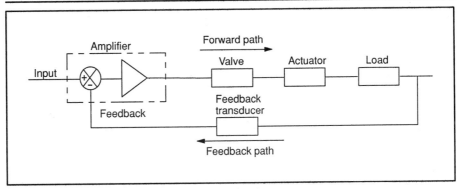

Figure 3.46

The FORWARD path includes the valve amplifier, control valve and actuator.

The FEEDBACK path includes the transducer and any associated signal conditioning devices.

Any errors that occur in the FORWARD path, (such as hysteresis, non-linearity, temperature drift, etc.) are minimized by the gain control of the amplifier, and can be eliminated altogether by use of an integrating amplifier (discussed in more detail later).

But in the FEEDBACK path, this is not the case. Any component errors in the feedback path will DIRECTLY affect the controlled variable via the error signal.

Therefore:

THE ACCURACY OF A CLOSED LOOP SYSTEM CAN NEVER BE ANY BETTER THAN THE ACCURACY OF THE FEEDBACK PATH, which includes the feedback transducer and associated signal conditioning devices.

In determining the accuracy of a closed loop system, transducer error must be added to any error caused by components in the forward path.

THIS ERROR CANNOT BE ELIMINATED BY MERELY USING A "MORE SOPHISTICATED" CONTROLLER.

Therefore, the correct selection and installation of the feedback transducer is vital to the performance of the system.

Although there are many different types of transducers which are made to measure a wide range of things, they can basically be categorized as either ANALOG or DIGITAL devices.

ANALOG transducers produce a continuous signal, usually a voltage or a current, which is somehow proportional to the sensed variable. One example of an analog transducer is the old gauge-type speedometer in cars. Vehicle speed is indicated by the position of a needle on a dial.

Ideally, the relationship between the sensed variable and the transducer signal is linear over the working range. Figure 3.47 shows typical outputs from both a linear and non-linear analog transducer.

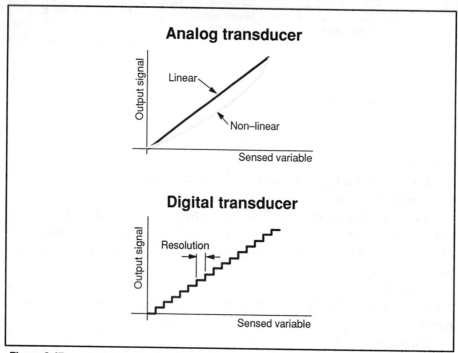

Figure 3.47

The control of systems and machinery is being performed increasingly by digital control devices, such as PLCs and microprocessors. In such cases, digital transducers are becoming popular because their signal outputs can be directly handled by these digital devices.

A digital signal is also shown above. It increases in definite steps, with the width of a step being equal to the RESOLUTION of the device. Resolution is the smallest unit that the digital device can measure accurately.

Analog transducers can be used in digital systems, and vice versa, if the transducer signal is first converted by an analog to digital or a digital to analog converter (Figure 3.48):

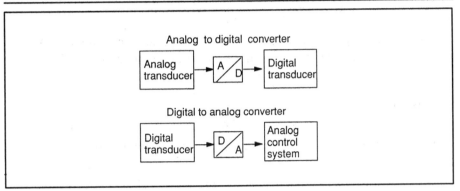

Figure 3.48

Digital position transducers (either linear or rotary) can be further divided into INCREMENTAL and ABSOLUTE types.

Incremental transducers produce a series of pulses to indicate position relative to some fixed point. The pulses are fed to an up/down counter, and by reading the number in the counter, the position relative to the fixed reference point can be determined; for example, the position of the transducer is now 23,782 pulses away from 0 inch position.

On startup, the counter must first be set to zero by physically moving the actuator to the zero-reference position. Having to zero out each actuator after, for example, a power failure may be a disadvantage in some applications, and may not be possible in others.

Absolute transducers are designed to overcome this problem by means of additional reference information built into the transducer. After its initial calibration, the transducer provides an absolute position indication and does not need to be re-referenced upon system re-start.

**Transducer
Considerations**

When selecting a suitable transducer for an application, you must take into account certain features of the transducer in order to avoid problems. These features are explained below.

1. LINEARITY ERROR

Linearity error (or non-linearity) is the maximum deviation of the actual transducer signal from the ideal "straight-line" path followed by a "perfect" transducer.

Linearity error is expressed as a percentage of maximum error over maximum rated output (Figure 3.49):

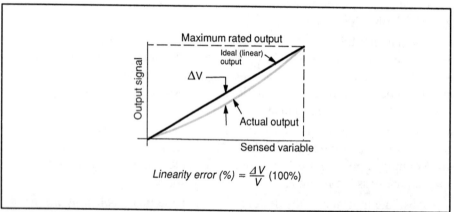

$$\text{Linearity error (\%)} = \frac{\Delta V}{V} (100\%)$$

Figure 3.49

For example, let's calculate the error you can expect from a pressure transducer and its associated electronics, which produces an output of 0 to 10 volts over its working range of 0 to 5800 PSI. Its quoted linearity is 0.5%.

Then, the maximum output error is 10V x .005 = 0.05 volts

This corresponds to a pressure error of:

$$\frac{5800 \; PSI}{10 \; volts} \times .05 \; volts = 29 \; PSI$$

2. HYSTERESIS

Hysteresis is the amount of deviation from the ideal output depending on whether the output signal is increasing or decreasing. It is quoted as a percentage of maximum rated output.

3. REPEATABILITY

Repeatability is the ability of a transducer to produce the same exact output for repeated applications of a given input. Repeatability and hysteresis errors do not take temperature variations into account.

4. TEMPERATURE DRIFT

The output of analog transducers can change with temperature. The temperature drift is expressed as a percentage of maximum rated output signal per degree of temperature change, over a specified range.

5. RESOLUTION

Digital transducers have finite levels of resolution, depending upon the number of signal "steps" over the range of the device. Resolution is also applicable in wire wound potentiometers. Resolution can be specified as a percentage of rated output, or as an absolute value (e.g., .0001 inch, .02 PSI, 5 lb_f, etc).

As a general rule, the resolution of the transducer should be AT LEAST 10 TIMES BETTER THAN THE ACCURACY REQUIREMENT OF THE SYSTEM.

For example, if a position control system needs to achieve an accuracy of 0.004 inch, then you need a transducer with a minimum resolution of 0.0004 inch.

6. RIPPLE

The electrical output of a transducer can sometimes include a certain amount of AC noise. This is especially true of transducers that use an AC voltage for their operation (inductive type transducers). It will be necessary to ensure that any ripple does not cause undesired feedback signals to the control system.

7. SPEED OF OPERATION

Many transducers have a limited speed of operation for either mechanical or electronic reasons. Maximum operating speed must be checked against the requirements of the application.

8. DYNAMIC RESPONSE

The frequency response of a transducer must also be compatible with the system being controlled.

AS A GENERAL RULE, THE NATURAL FREQUENCY OF THE TRANSDUCER MUST BE AT LEAST 10 TIMES GREATER THAN THAT OF THE CONTROLLED SYSTEM.

9. INSTALLATION

Careful consideration must be given to the mechanical installation of the transducer. This will directly influence the accuracy and performance of the system.

Transducers must be installed according to the manufacturer's recommendations. This usually means that they must be rigidly affixed and free of play or backlash.

The transducer should be either designed for the environment it is to operate within, or protected by an appropriate enclosure. All electrical cables must be properly shielded if electrical interference is possible.

10. LIFE EXPECTANCY

All mechanical contact transducers (e.g. potentiometers), have a limited life expectancy or cycle life. It is typically rated in thousands of cycles or millions of actuations. This may be very critical to an application where reliability is important. It can also be a useful predictor in stocking spare parts for your system.

Typical Transducer Operation

The variables that most commonly require sensing in a closed loop electrohydraulic system are:

- Linear Position
- Rotary Position
- Linear Velocity
- Rotary Velocity
- Pressure
- Force
- Torque

Linear Position Transducers:

The simplest type of linear position transducer is the linear potentiometer. It consists of a carbon or conductive plastic strip placed inside of a housing. A supply voltage is applied across the strip, and a wiper is moved along the strip by the actuating rod (Figure 3.50):

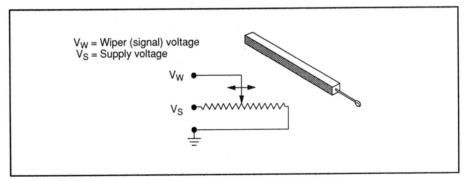

V_W = Wiper (signal) voltage
V_S = Supply voltage

V_W

V_S

Figure 3.50

Linear pots provide good linearity, and the use of the conductive strip means that resolution is virtually infinite. However, the fact that mechanical contact exists between the wiper and the strip results in wear, which gives the device a limited life expectancy, and a relatively low frequency range (typically less than 5 Hz).

Another type of linear position transducer is the LVDT (short for LINEAR VARIABLE DIFFERENTIAL TRANSFORMER) shown in Figure 3.51:

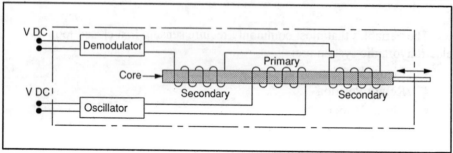

Figure 3.51

The LVDT consists of a primary coil and two secondary coils that surround a soft iron core. The core is connected to whatever we are measuring the position of (actuator rod, valve spool, etc.)

A high frequency (1000 Hz) AC signal is fed to the primary coil, which creates a magnetic field in the core. This AC signal is usually produced by a DC device known as an "oscillator," which can be built right into the housing of the LVDT.

This magnetic field induces a voltage in the secondary coils, like in a transformer. The two coils are connected in opposition.

If the core is exactly centered, the voltages induced in the secondaries cancel each other out, and produce a zero output. However, if the core is moved even slightly away from center, the voltage induced in one secondary will increase, with a corresponding reduction in the other secondary. This produces a voltage difference which is proportional to the amount of movement.

The phase relationship in this voltage difference indicates the direction of movement.

The output from the secondaries is fed to a "demodulator," which converts the AC signal to a DC voltage. In other words, the demodulator performs signal conditioning on the LVDT output, to convert it to a DC voltage that can be used by the amplifier module. In many LVDTs, the demodulator is also built into the LVDT housing.

The LVDT typically has a long cycle life, because the iron core does not contact the coils and is very resistant to wear. It also has a high frequency range. However, due to an effect called "zero-crossover error" which is commonly found in AC devices, the linearity of an LVDT is typically not as good as that of the linear pot.

It is becoming increasingly common to install the position transducer directly inside of a cylinder, both for convenience and protection of the transducer. One linear transducer that is commonly installed this way is referred to as a MAGNETOSTRICTIVE type transducer (Figure 3.52):

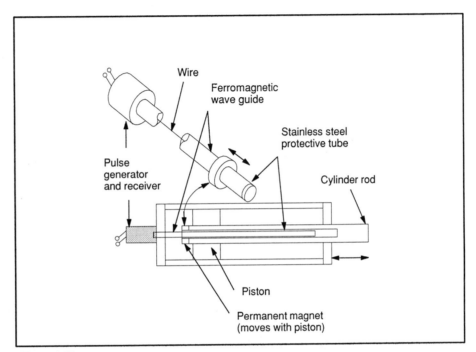

Figure 3.52

The cylinder has a hollow rod, to allow insertion of the transducer probe. A doughnut shaped magnet is affixed to the back of the piston. The probe goes through the center of the magnet and into the hollow cylinder rod. As the piston moves, it carries the magnet along the probe.

The probe consists of a tube made of a magnetostrictive alloy (an alloy in which strain is produced in the presence of a magnetic field). Inside of the tube is a loop of wire.

An electrical pulse is sent down the wire. When the pulse meets the magnetic field of the doughnut shaped magnet, a strain pulse is produced in the tube, which travels down the tube to the Pulse Generator/receiver head and is sensed there.

By measuring the amount of time elapsed between sending the electrical pulse down the wire and receiving the strain pulse back, the position of the piston can be very accurately determined.

This device also has a long cycle life, because there is no physical contact between the probe and the actuator.

Rotary Position Transducers:

As in linear applications, potentiometers can be used for rotary situations, but are subject to the same mechanical wear and damage problems.

The rotary equivalent of an LVDT is the RVDT (Rotary Variable Differential Transformer). Its operating principle is the same as that of an LVDT, except that instead of a soft iron core, it uses a specially shaped cam. This provides a non-contact rotary position measuring device.

Another non-contact transducer is the optical encoder. It is available in either linear or rotary form, and can be either incremental or absolute type.

In a typical incremental shaft encoder, a glass disc with small, evenly spaced radial lines printed on its outer edge is rotated between an LED (light emitting diode) and a phototransistor receiver which senses presence of light (Figure 3.53):

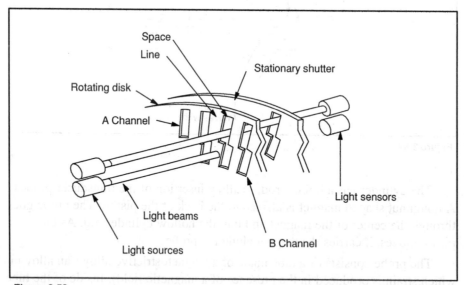

Figure 3.53

A stationary disc, called a mask, is installed close to the rotating disc to create a shutter mechanism.

The lines and spaces on the rotating disc alternately interrupt the beam of light from the LED, or allow it to pass through. This causes the phototransistor to generate a series of pulses, where each pulse represents a certain amount of rotation. The greater the number of lines and spaces, the finer the resolution.

If necessary, the output of the encoder can be converted from digital pulses to an analog signal.

Rotary Velocity Transducers:

Angular velocity is measured by a device called the tachometer-generator (or tach-gen, for short).

This device is basically a permanent magnet type DC generator (Figure 3.54):

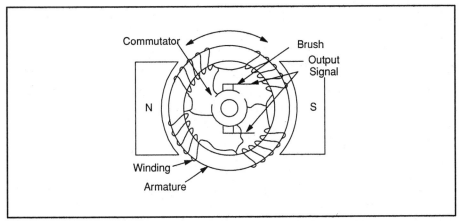

Figure 3.54

The tach-gen produces a DC output which is proportional to the speed of rotation of its shaft. It is almost unique among transducers in that it generates its own output voltage and does not require a power supply.

This transducer can also be used with a rack and pinion arrangement to convert angular velocity to linear velocity.

Actuator velocity can also be sensed indirectly by sensing the flow rate through the actuator. A typical flow transducer is shown in Figure 3.55:

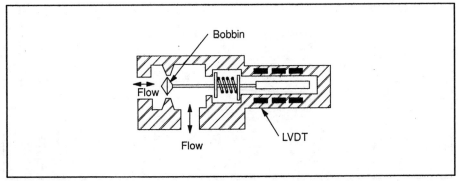

Figure 3.55

A spring loaded bobbin is located in the throat of the flow sensor. An LVDT is attached to the bobbin spindle. As flow passes through the transducer in either direction, the bobbin is displaced by an amount proportional to the flow through the device. The bobbin and throat are specially shaped to provide a near-linear relationship between flow rate and bobbin movement. Since bobbin movement is sensed by the LVDT, an electrical signal proportional to flow rate is obtained.

The main limitations of such devices are the comparatively large linearity errors, and the fact that they produce unreliable readings at low flow rates.

Pressure, Force and Torque Transducers:

In many closed loop force or torque control applications, it may be more convenient to sense actuator pressure rather than to directly sense force or torque.

Pressure transducers normally operate on either a strain gage or a piezo electric principle.

When used with the necessary signal conditioning electronics, these devices produce a voltage or current that is proportional to pressure. And since these types of devices contain no moving parts, they are normally very reliable in operation.

Differential pressure types are also available to sense pressure difference across actuator ports.

Torque and force transducers (sometimes called load cells) also operate on the strain gage principle and are available in many different sizes and styles to suit almost any application (Figure 3.56):

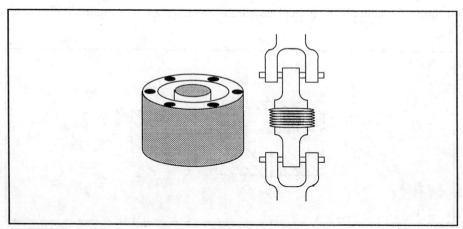

Figure 3.56

Valve Sizing

Traditionally, servo valve flow ratings are quoted assuming a 500 PSI pressure drop per flow path, or a total of 1000 PSI across the valve.

Unfortunately, few systems happen to create exactly 1000 PSI of pressure drop across the valve. Therefore, it is necessary to determine the size of servo valve, based on the requirements of the system that the valve will be controlling.

The starting point for valve sizing is the actuator, since the work to be done by the actuator is the reason that the system exists.

The actuator is normally a cylinder or a motor. Let's start with a cylinder application.

Cylinder Applications

Consider the servo system shown in Figure 4.1:

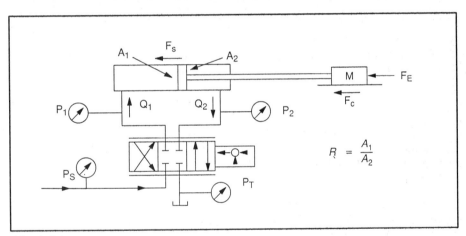

Figure 4.1

To correctly size the valve for this UNEQUAL AREA CYLINDER, we'll have to look at the system during extension AND retraction.

Let's start with the case where the cylinder is extending (pushing the load). The load is resting on a smooth floor.

To push the load, we must overcome several different forces:

- F_a, which is the force required to accelerate the load.
- F_C, which is the force required to overcome the floor's friction against the load.
- F_S, which is the force required to overcome the friction of the seals in the cylinder itself.
- F_E, which is any other external force happening to act on the load (Examples: the wind, the force of a cutting tool moving across the load, dirt on the smooth floor, etc.)

In order to move the load, enough pressure must be applied across the cylinder ports to overcome these forces. Our required force, then is:

$$F = F_a + F_C + F_S + F_E$$

DETERMINING THE LOAD ACCELERATION FORCE:

F_a is the force required to accelerate the load. If you recall from your High School physics lessons,

$$F = M \times a$$

or

$$Force = Mass \times acceleration$$

Also recall that the WEIGHT of an object is actually a FORCE.

$$Weight = Force = Mass \times acceleration$$

On earth, the acceleration that holds us on the surface of the earth is gravity (g). Gravity is an acceleration of 32.2 ft/sec^2. Therefore,

$$Weight = Force = M \times g$$

On the moon, g is about 1/6 of earth's gravity. If a 180 lb man went to the moon, he would only weigh about 30 lbs. there. HOWEVER, HIS MASS IS THE SAME ON EARTH, ON THE MOON, OR IN THE WEIGHTLESSNESS OF SPACE!

The easiest way to determine the MASS of our load, is to divide its weight by earth's gravity (g). Let's refer to the load's weight as W_L:

If

$$W_L = M \times g$$

then

$$M = \frac{W_L}{g}$$

We can now take the mass and multiply it by our required acceleration to determine the load acceleration force, F_a.

Acceleration can be determined by taking the maximum load velocity and dividing it by the time available for acceleration. For example, if we require our cylinder to go from a speed of 0 to 15 inches/sec within .25 seconds after opening the valve, then we need an acceleration of

$$a = \frac{Velocity}{Acceleration\ time} = \frac{15\ in/sec}{.25\ sec} = 60\ in/sec^2$$

If we know the weight of the load, the maximum velocity needed, and how fast we must accelerate the load, then we can readily calculate the force required to accomplish this as:

$$F_a = \left(\frac{W_L}{g}\right) \times a$$

DETERMINING THE LOAD FRICTIONAL FORCE:

F_C is the force required to slide the load across the floor. It is equal to the weight of the load times the friction coefficient of the floor.

The friction coefficient, μ, is a unitless number which indicates how easily the load will slide across the floor. The bigger the value of μ, the more friction there is.

If the load is level on the floor, then:

$$F_C = \mu \times W_L$$

(Note: If the load is on an incline, we must consider the angle of incline as well. We'll keep this example simple. If you are unfamiliar with the FRICTION COEFFICIENT, it is suggested that you refer to a standard Physics text under the subjects of Mechanics, and Static and Kinetic Friction.)

DETERMINING THE CYLINDER SEAL FRICTION:

F_S is the amount of force required to move the unloaded cylinder itself. This information is usually available in the catalog data for the cylinder. If this information is not available readily, you can assume it to be about 10% of the total force requirement, or:

$$F_S = 0.10F$$

TOTAL FORCE REQUIRED FOR CYLINDER EXTENSION:

F is now the sum of all of these individual forces:

$$F = \frac{W_L\,(a)}{(32.2)\ 12} + F_C + F_E + F_S \ \ lb_f$$

Equation #5

Where: F = Total Cylinder Extension Force (lb_f)
  W_L = Load Weight (lb_f)
  a = Required Load Acceleration (in/sec^2)

F_C = Load Friction (lb$_f$)
F_E = External Forces (lb$_f$)
F_S = Seal Friction Force (lb$_f$)
32.2 = Acceleration due to Gravity (ft/sec^2)
12 = Conversion Factor for feet to inches

There is yet another force to be considered in this process.

When oil is ported to the cap end of the cylinder and the load begins to move, we will have BACKPRESSURE in the rod end of the cylinder, since oil is being pushed out to the tank through the other side of the valve.

We therefore need to FORCE BALANCE this cylinder, as a step in figuring out WHAT THE FLOW RATING OF THE VALVE NEEDS TO BE.

The Force applied to the cap end must equal the backpressure force, PLUS the force required to move the load:

$$F_{cap} = F_{rod} + F$$

Recall from your training in Basic Hydraulics that:

$$Force = Pressure \times Area$$

Therefore,

$$F_{cap} = P_1 \times A_1$$
(A$_1$ = cap end area)

and

$$F_{rod} = P_2 \times A_2$$
(A$_2$ = rod end area)

The force balance equation during cylinder extension then becomes:

$$P_1 A_1 = P_2 A_2 + F$$

Since the cylinder shown is an UNEQUAL area actuator, the flow through the two sides of the servo valve will NOT be the same:

$$Q_1 = Q_2 \ (R)$$

where "R" is the area ratio of the cylinder, that is:

$$R = \frac{A_1}{A_2}$$

Now, recall from Equation #1 (page 64) that FLOW through a servo valve is PROPORTIONAL TO THE SQUARE ROOT OF THE PRESSURE DROP.

This means that:

$$Q_1 :: \sqrt{P_S - P_1}$$

and

$$Q_2 :: \sqrt{P_2 - P_T}$$

Since $Q_1 = Q_2$ (R), you can now make the claim that:

$$\sqrt{P_S - P_1} \;=\; \sqrt{P_2 - P_T} \;\; (R)$$

If you SQUARE both sides of the equation, you now have a solid mathematical relationship between P1 and P2 that you can use:

$$(P_S - P_1) \;=\; (P_2 - P_T)(R^2)$$

With a little algebra, you can manipulate the above equations (See Appendix D) and come up with the following expressions:

$$P_1 \;=\; \frac{P_S(A_2) \;+\; R^2[F \;+\; P_T(A_2)]}{A_2(1 + R^3)}$$

Equation #6

AND

$$P_2 \;=\; P_T \;+\; \left[\frac{P_S - P_1}{R^2}\right]$$

Equation #7

Once we know the values of P_1 and P_2, we know the pressure drop across the valve, and we can calculate the necessary flow rating of the valve.

The actual flow needed through the valve is determined by the size of the actuator and the actuator speed required:

$$Q_A \;=\; \frac{V_{max}(A_1)60}{231} \;\; gpm$$

Equation #8

Where: Q_A = Actual Flow Required through valve (GPM)

V_{max} = Maximum Required Actuator Velocity (in/sec)

A_1 = Cap end area (in^2)

60 = Conversion factor – seconds to minutes

231 = Conversion factor – cubic inches to gallons

Once the actual flow and the pressure drops are determined, we can now figure out what flow rating is needed for cylinder extension.

Equation #2 (page 65) provides the information we need. As shown, it calculates actual flow through a given size of valve at a known pressure drop.

What we need to know is the RATED flow, since we already know the ACTUAL flow and the pressure drop.

Let's turn Equation #2 around and assume 100% input signal is being applied.

Let's also modify the equation to give us the pressure drop only across ONE OF THE TWO FLOW PATHS THROUGH THE VALVE – the flow path leading to the cap end of the cylinder, which is the one that will be passing the larger flow.

Equation #2 then becomes:

$$Q_R = Q_A \sqrt{\frac{500}{P_S - P_1}} \quad gpm$$

Equation #9

THE RETRACTION STROKE:

We must also analyze what happens to flow during the RETRACTING stroke. The valve will shift, porting Pressure to the Rod end and Tank to Cap end.

If we carry out the analysis for retraction, WE WILL FIND THAT OUR EQUATIONS CHANGE TO:

$$P_1 = P_T + (P_S - P_2)R^2$$

Equation #10

$$P_2 = \frac{P_S(A_2)R^3 + F + P_T(A_2)R}{A_2(1 + R^3)}$$

Equation #11

$$Q_A = \frac{V_{max}(A_2)60}{231} \; gpm$$

Equation #12

$$Q_R = Q_A \sqrt{\frac{500}{P_S - P_2}} \; gpm$$

Equation #13

Using these equations, we can determine the flow rating required for cylinder retraction. THE SIZE OF THE VALVE WILL BE DETERMINED BY WHICHEVER VALUE OF Q_R IS GREATER (EXTENDING OR RETRACTING).

Example 4.1:

Determine the rated flow of the servo valve to be used in the below system.

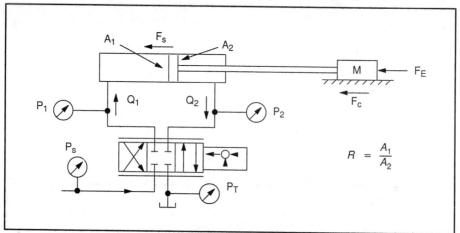

Figure 4.2

Where: P_S = 3000 PSI $\quad$ V_{max} = 15 in/sec
$\qquad P_T$ = 75 PSI $\qquad$ W_L = 2645 lb_f
$\qquad A_1$ = 8.26 in^2 $\qquad$ Acceleration Time = 25 msec
$\qquad A_2$ = 5.9 in^2 $\qquad$ μ = 0.32
$\qquad F_E$ = 3930 lb_f $\qquad$ F_S = 900 lb_f

1. The acceleration of the load is V_{max}/Accel Time

$$a = \frac{15 in/sec}{0.025 sec} = 600 in/sec^2$$

2. The load acceleration force is

$$F_a = \frac{W_L(a)}{32.2 \times 12} = \frac{2645 \ lb_f \times 600 \ in/sec^2}{32.2 \ ft/sec^2 \times 12 \ in/ft} = 4107 \ lb_f$$

3. The load friction force is

$$F_C = W_L \times \mu = 2645 \ lb_f \times 0.32 = 846.4 \ lb_f$$

4. Total Force F can now be determined:

$$F = F_a + F_C + F_E + F_S$$
$$F = 4107 + 846.4 + 3930 + 900 = 9783.4 \ lb_f$$

P_S = 3000 PSI

P_T = 75 PSI

A_1 = 8.26 in^2

A_2 = 5.9 in^2

F_E = 3930 lb$_f$

F = 9783.4 lb$_f$

V_{max} = 15 in/sec

W_L = 2645 lb$_f$

Acceleration Time = 25 msec

μ = 0.32

F_S = 900 lb$_f$

5. Calculate the cylinder area ratio:

$$R = \frac{A_1}{A_2} = \frac{8.26}{5.90} = 1.4$$

6. Now calculate P_1 and P_2:

$$P_1 = \frac{P_S(A_2) + R^2[F + P_T(A_2)]}{A_2(1 + R^3)}$$

$$P_1 = \frac{3000(5.9) + 1.4^2\ [9783.4 + 75(5.9)]}{5.9(1 + 1.4^3)}$$

P_1 = 1708.6 PSI

$$P_2 = P_T + \left[\frac{P_S - P_1}{R^2}\right]$$

$$P_2 = 75 + \frac{(3000 - 1708.6)}{1.4^2}$$

P_2 = 733.9 PSI

7. Our problem requires the cylinder to accelerate to a maximum speed of 15 in/sec. Therefore, we need to supply the following flow to the cap end of the cylinder:

$$Q_A = \frac{V_{max}(A_1)60}{231}\ gpm$$

$$Q_A = \frac{15\ in/sec \times 8.26\ in^2 \times 60\ sec/min}{231\ in^3/gallon}$$

Q_A = 32.18 PSI

We need 32.18 GPM to move the cylinder at a speed of 15 in/sec.

$P_S = 3000$ PSI

$P_T = 75$ PSI

$A_1 = 8.26$ in^2

$A_2 = 5.9$ in^2

$F_E = 3930$ lb$_f$

$F = 9783.4$ lb$_f$

$V_{max} = 15$ in/sec

$W_L = 2645$ lb$_f$

Acceleration Time = 25 msec

$\mu = 0.32$

$F_S = 900$ lb$_f$

8. Do we need a 32 GPM valve for this application ? NO!
    A 32 GPM valve passes 32 GPM at 500 PSI drop per flow path.
    But this application produces a P to A pressure drop of:

    $$P_S - P_1 \quad = \quad 3000 - 1708.6 \quad = \quad 1291.4\,PSI$$

    Therefore, we need to DERATE the valve we wish to use.

    $$Q_R \;=\; Q_A \sqrt{\frac{500}{P_S - P_1}} \;\; gpm$$

    $Q_R = 32.18 \sqrt{\dfrac{500}{1291.4}}$

    $Q_R = 20$ GPM

    WE NEED A 20 GPM VALVE TO EXTEND THE CYLINDER AS SPECIFIED.

9. Now let's analyze the RETRACTING STROKE. Assume that the velocity and acceleration requirements are the same, and that F_E still acts on the load.

    $$P_2 \;=\; \frac{P_S(A_2)R^3 \;+\; F \;+\; P_T(A_2)R}{A_2(1 + R^3)}$$

    $P_2 = \dfrac{3000(5.9)(1.4)^3 + 9783.4 + 75(5.9)1.4}{5.9(1 + 1.4^3)}$

    $P_2 = 2669.66$ PSI

    $$P_1 \;=\; P_T + (P_S - P_2)R^2$$

    $P_1 = 75 + (3000 - 2669.66)1.4^2$

    $P_2 = 722.47$ PSI

P_S = 3000 PSI

P_T = 75 PSI

A_1 = 8.26 in^2

A_2 = 5.9 in^2

F_E = 3930 lb$_f$

F = 9783.4 lb$_f$

V_{max} = 15 in/sec

W_L = 2645 lb$_f$

Acceleration Time = 25 msec

μ = 0.32

F_S = 900 lb$_f$

$$Q_A = \frac{V_{max}(A_2)60}{231} \ gpm$$

$$Q_A = \frac{15 \ (5.9) \ 60}{231}$$

Q_A = 23 GPM

$$Q_R = Q_A \sqrt{\frac{500}{P_S - P_2}} \ gpm$$

$$Q_R = 23 \sqrt{\frac{500}{3000 - 2669.66}}$$

Q_R = 28.3 GPM

A SURPRISING SITUATION OCCURS HERE! Since it takes less flow to retract the cylinder than to extend it, you would intuitively expect the flow rating to be lower for retraction. But we find that the opposite is true!

Compare the data:

	Extending	Retracting
Actual Flow Required	32.2 GPM	23.0 GPM
Rated Flow of Valve	20.0 GPM	28.3 GPM
Pressure Drop Across	1291.4 PSI	330.3 PSI
Active Flow Path	(P to A)	(P to B)

Notice that during extension, the pressure drop is higher than 500 PSI across the active flow path. This high pressure drop will push more oil through the valve – therefore, we can use a smaller valve.

However during retraction, the pressure drop across the active path is lower than 500 PSI, meaning that it will take a larger valve to pass the needed 23 GPM during retraction.

This example therefore requires a valve sized to handle the larger flow requirement, which occurs during cylinder retraction. We will need a valve rated for 28.3 GPM minimum.

Motor Applications

We can analyze motors in a similar way to cylinders, taking into account some differences.

In a cylinder, we deal with forces whereas, in a motor we are concerned with TORQUES.

Motors are typically regarded as EQUAL AREA ACTUATORS, meaning that the results calculated for forward revolution of the motor tend to also apply to reverse revolution.

Consider the system shown in Figure 4.3:

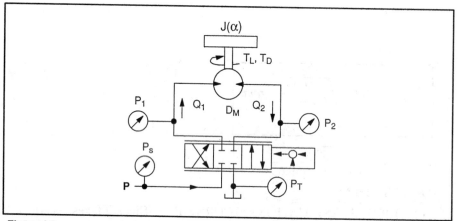

Figure 4.3

Figure 4.3 shows a servo valve controlling a hydraulic motor.

The motor has a certain displacement, D_M, which is expressed in cubic inches per revolution (in³/rev).

The load that is turned by the motor will have a certain amount of "rotational inertia." Rotational inertia (J) is determined by the mass, size and geometry of the load. The greater the rotational inertia of the load, the more force will be required to accelerate the load.

The total torque required to turn the motor is made up of three major parts:

- T_a, the torque required to ACCELERATE the load. This is derived from the expression $T_a = J(\alpha)$

    where:

    T_a = acceleration torque (lb$_f$ in)

    J = total motor shaft inertia, including load and rotating parts of motor (in lb$_f$ sec²)

    α = angular acceleration (radians/sec²)

- T_L, the torque required to KEEP THE LOAD MOVING (lb_f in)
- T_D, the damping torque of the motor, which is the torque required to overcome friction and drag in the motor itself (lb_f in)

The TOTAL TORQUE required for the system is T:

$$T \ = \ J(\alpha) + T_L + T_D \ \ lb_f \ in$$

Equation #14

If we were to do a complete analysis on motors, as we did on cylinders, we would eventually come up with the following ways to find P_1 and P_2:

$$P_1 \ = \ \left[\frac{P_S + P_T}{2} \right] + \left[\frac{\pi \ T}{D_M} \right] \ psi$$

Equation #15

and

$$P_2 \ = \ P_S - P_1 + P_T \ \ psi$$

Equation #16

After establishing what the pressure drops will be in our system, we can determine the actual flow rate required using:

$$Q_A \ = \ \frac{N(D_M)}{60} \ \ in^3/sec$$

or

$$Q_A \ = \ \frac{N(D_M)}{231} \ \ gpm$$

Equation #17

Once we know the pressure drops and the actual flow rate needed, we can determine the size of the servo valve required using the same derating equation as before:

$$Q_R = Q_A \sqrt{\frac{500}{P_S - P_1}} \ gpm$$

Equation #9

If the load changes for the two directions of rotation, Q_R should be evaluated for both cases, and the larger value should be used to select the valve.

The next page contains an example of sizing a servo valve for a motor application.

Example 4.2:

Determine the size of servo valve required for the system shown in Figure 4.4:

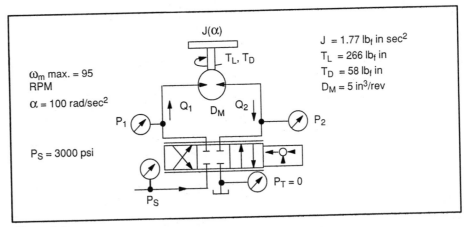

Figure 4.4

1. $T_a = J(\alpha) = 1.77\ lb_f\ in\ sec^2 \times 100\ radians/sec^2$

   $T_a = 177\ lb_f\ in$

2.
   $T = T_a + T_D + T_L = 177 + 58 + 266\ lb_f\ in = 501\ lb_f\ in$

3. $P_1 = \dfrac{P_S + P_T}{2} + \dfrac{\pi\ T}{D_M} = \dfrac{3000 + 0}{2} + \dfrac{3.1417\ (501)}{5}$

   $P_1 = 1500 + 314.79 = 1814.79\ PSI$

4. $P_2 = P_S - P_1 + P_T = 3000 - 1814.79 + 0 = 1185.21\ PSI$

5. $Q_A = \dfrac{N\ D_M}{231} = \dfrac{95\ (5)}{231} = 2.06\ GPM$

6. $Q_R = Q_A \sqrt{\dfrac{500}{P_S - P_1}} = 2.06 \sqrt{\dfrac{500}{3000 - 1814.79}}$

   $Q_R = 1.34\ GPM$

Hydraulic System Considerations

When using servo valves or high performance proportional valves to control a hydraulic device, there are two basic types of hydraulic systems which can be considered:

Servo actuators or Servo Pumps.

We have so far assumed that the control valve is directly controlling the hydraulic actuator (cylinder, motor, etc.), and that the valve is provided with a constant supply pressure and flow from the hydraulic power unit.

But it is also possible to control the actuator indirectly, by using the valve to control the flow from a variable displacement pump.

Servo Actuator Control

In a servo actuator, the valve is mounted near to or upon the actuator, and it directly controls the flow to the actuator ports (Figure 5.1):

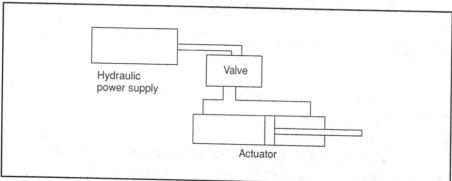

Figure 5.1

One of the major factors in determining the performance of a closed loop system is the "STIFFNESS" of the actuator & load assembly.

Increasing the stiffness enables us to increase the maximum system gain that can be used, resulting in faster response and greater accuracy.

STIFFNESS of the system is affected by the volume of hydraulic fluid that is under compression. The less oil we have under compression, the greater is the stiffness of the system.

By mounting the valve as close to the actuator as possible, the volume of compressed oil can be held to a minimum, maximizing system stiffness. Servo valves are normally mounted DIRECTLY UPON the actuator ports to keep the oil volume to an absolute minimum.

A servo valve requires a constant pressure supply in order to give consistent and predictable performance. The simplest way of providing this is by using a fixed displacement pump and a relief valve (Figure 5.2):

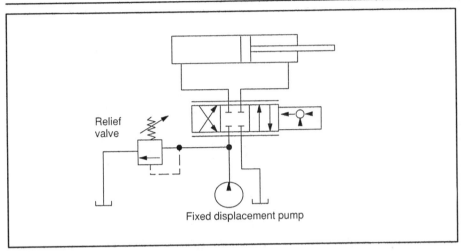

Figure 5.2

One problem with this simple arrangement is its inefficiency. Any flow which is not used by the valve will pass over the relief valve at the full relief valve pressure setting, creating heat.

This power waste can be especially troublesome in position or pressure control applications, where the actuator does not move for long periods of time.

If system pressure must be maintained in order to hold the actuator in position, then it will not be practical to simply unload the relief valve.

One method of unloading the pump to reduce heat generation during no-flow periods, while maintaining load-holding pressure to the valve, is shown in Figure 5.3:

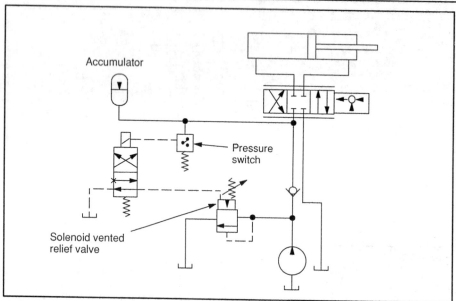

Figure 5.3

Flow passes from the pump to the servo valve, at a pressure determined by the setting of the relief valve.

This flow also pressurizes the accumulator.

System pressure is sensed by the pressure switch. When system pressure reaches the setting of the switch, the switch energizes the solenoid valve.

The solenoid valve connects the vent of the relief valve to tank, causing the relief valve to open and unload the pump.

The accumulator now maintains holding pressure to the servo valve. The check valve prevents the accumulator from discharging through the relief valve.

This system will have a certain amount of pressure variation, due to the deadband of the pressure switch. For example, the pump may unload at 1000 PSI, and "kick-on" again at 750 PSI.

The pressure variations can be minimized, at additional cost, by using a larger accumulator and a pressure switch with a smaller deadband.

Another method of supplying a constant supply pressure to a servo valve, without creating large amounts of waste heat, is through the use of a pressure compensated variable pump (Figure 5.4):

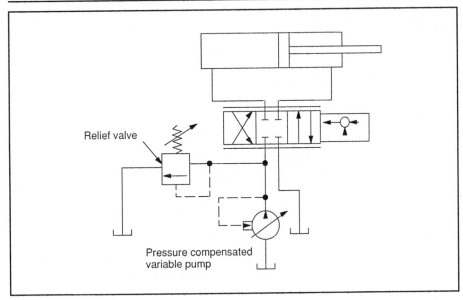

Relief valve

Pressure compensated
variable pump

Figure 5.4

When the servo valve is operating at or near null position, the pump flow automatically reduces to near zero.

The amount of pressure variation between full pump flow and zero flow will be relatively small, since this is now determined by the pressure override setting of the pump's compensator valve (typically about 75 PSI).

In systems with large pumps or long pipe runs between pump and actuator, the pressure compensated variable pump can take a relatively long time to respond to a large change in flow demand. This is a disadvantage only in fast acting systems where response speed is critical.

Even then, the problem can be addressed by installing an accumulator close to the servo valve (Figure 5.5) which will maintain system pressure to the valve during the time lag between valve opening and pump response to the increased demand.

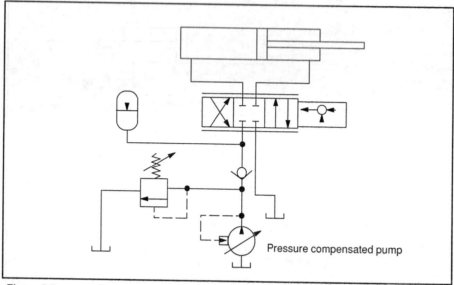

Figure 5.5

Pressure fluctuations and low system pressure in a servo system have the most drastic effect on the pilot stage of the servo valve (the flapper/nozzle area). One very effective way to overcome such problems is to supply the valve with a separate pilot pressure supply as shown in Figure 5.6:

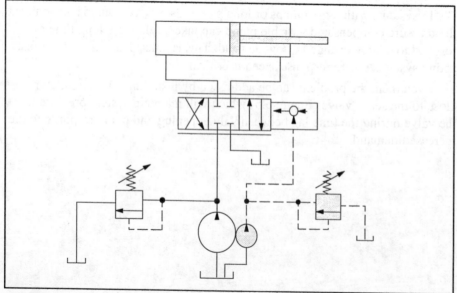

Figure 5.6

An additional small pump (about .5 GPM) and relief valve can be added to provide a dedicated pilot pressure to the servo valve, which will be unaffected by conditions in the rest of the system.

Whatever option is used, the servo or proportional valve will still introduce some inefficiency into the system, in the form of heat, due to the relatively high pressure drops across the valve.

The heat created by the valve pressure drop must be considered in the design of the system, to avoid failure of the hydraulic fluid and subsequent problems with the hydraulic system.

The problem of excess heat generation will generally be more severe in velocity control systems or rapid cycling applications, where the valve tends to remain in the operated (fully or partially open) position for long periods of time.

Heat generation may not be a real problem in position or pressure control applications, where the valve remains closed most of the time.

Servo Controlled Pump

A servo valve can be used with a variable displacement pump to control the angle of a swash plate or cam ring (Figure 5.7), thereby controlling the system flow rate.

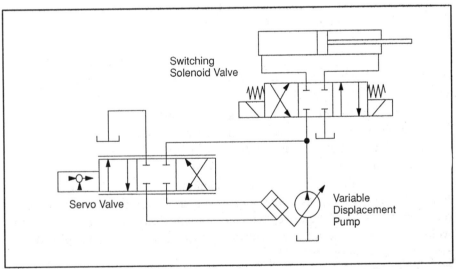

Figure 5.7

The main advantage of servo controlled pumps is the increase in system efficiency which can be obtained. The pump produces only the amount of flow required at a system pressure equal to load pressure.

In our example above, flow is transmitted through a solenoid valve to the actuator. Since this valve is fully open or fully closed, it has a very low pressure drop across it, resulting in very little heat generation and power loss at the solenoid valve.

The servo valve which is controlling the pump will be a small, low flow valve since it only has to be big enough to operate the pump's control piston.

It does not have to be sized to handle full system flow.

The main disadvantage of a servo pump is that there is usually a large volume of oil between the pump and the actuator, resulting in lowered system stiffness. This will result in slower system response to changes and a lower degree of accuracy.

Servo pump systems are normally used in high power transmission applications such as mobile vehicles, where high system efficiency is more important than system response time or accuracy.

Filtration

The servo valves and high performance proportional valves we have discussed are precision made devices which use matched sliding components and small, sharp edged metering orifices which **must** be protected from contamination particles. Suitable filtration is NOT an optional choice in such systems. It is an absolute necessity if reliable operation is expected.

Non-bypass type, pressure-line filtration is recommended for servo systems. If the environment is especially harsh, off-line filtration may also be necessary.

An ISO cleanliness level of 15/13/10 is recommended for servo systems. This ISO level means that within each 1 milliliter of your system fluid, there should be:

fewer than 320 particles greater than 2 microns in size, and
fewer than 80 particles greater than 5 microns in size, and
fewer than 10 particles greater than 15 microns.
(1 micron = .000001 meter = .0000394 inch)

Before installing one of these close-clearance valves on a new or a refurbished system, it is essential to carry out an effective flushing procedure on the entire system to remove "built-in" contamination particles (for example: weld scale, rust, dust, lint, metal shavings, blast-sand, paint chips, pipe dope, etc.)

During flushing, the servo or proportional valve is removed and replaced with an inexpensive flushing module. This module permits unrestricted flow through the system and can be manually shifted to allow flow in either direction during flushing of the system.

The pressure ratings of these flushing modules are usually limited to a value considerably lower than normal system operating pressure. Care should be taken to adjust system pressure to the manufacturers recommended flushing pressure.

Flushing may take several hours and several filter elements, until the necessary level of cleanliness has been achieved. For more detailed information on proper filtration procedures, refer to Vickers Publication #561 "Systemic Contamination Control."

A complete textbook on this topic is also available, in publication 5047.00/EN/1196/T entitled "Bird Bones & Sludge".

Closed Loop
System Analysis

The accurate prediction of how a closed loop system will perform under all circumstances requires an engineering-level knowledge of control systems theory. In the average high-performance industrial machine, the combined skills of electrical, electronic, mechanical, hydraulic and industrial engineers are employed, and they use some fairly sophisticated mathematical tools to design these machines.

Obviously, this single text cannot provide this level of understanding.

Therefore, this chapter provides a practical method for ESTIMATING the performance of a closed loop system. Some simplifications and general assumptions have been made which will tend to give conservative results.

If the estimated level of performance does not meet specifications, this does NOT mean that the specification cannot be met. It does mean that a more detailed analysis or a more sophisticated control technique may be required.

Block Diagrams

One way that we can simplify a Closed Loop System, is to view each important component, or group of components, as a block.

Each block has an INPUT and an OUTPUT.

The block reacts to the INPUT in some way, in order to produce the OUTPUT.

The relationship between the input and the output is referred to as the block's TRANSFER FUNCTION, or GAIN (G).

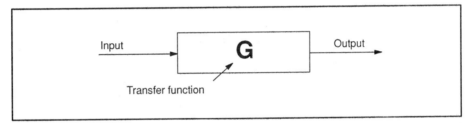

Figure 6.1

The GAIN of the block shown above is:

$$Gain \; = \; G \; = \; \frac{Output}{Input}$$

Equation #18

For example, suppose that the box shown above represents a 4 to 1 reduction gearbox (Figure 6.2).

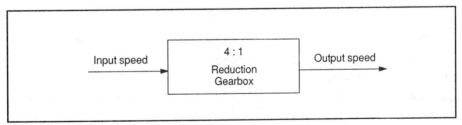

Figure 6.2

A 4:1 reduction gear causes the output shaft to turn once for every 4 turns of the input shaft. Therefore, the TRANSFER FUNCTION of the gear box is:

$$Gain = G = \frac{Output}{Input} = \frac{1}{4} = 0.25$$

Since the input and output are both SPEED, the transfer function is just a multiplication factor. For example, if the input speed were 500 RPM, the output speed would be:

$$500 \, RPM \times 0.25 = 125 \, RPM$$

This does not always apply. For example, let's look at a block representing a hydraulic pump (Figure 6.3):

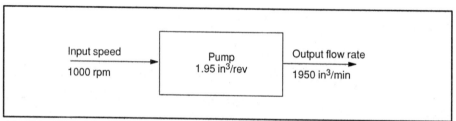

Figure 6.3

The input to the pump is the drive SPEED of the shaft (in RPM).

The output is a flow rate (in cubic inches per minute).

The transfer function is:

$$G = \frac{Output}{Input} = \frac{1950 \;\; in^3/min}{1000 \;\; rev/min} = 1.95 \;\; in^3/rev$$

The transfer function does not give a math model of the component under all possible dynamic conditions.

For example, if we apply 100,000 RPM to our pump, we probably won't get 195,000 in³/min (844 GPM) out of it. This is far beyond the pump's capabilities.

Therefore, it is important to apply common sense to the use of transfer functions, since they assume that the component is operating under steady conditions and within its design capabilities.

Now let's consider a simple closed loop system as shown in Figure 6.4:

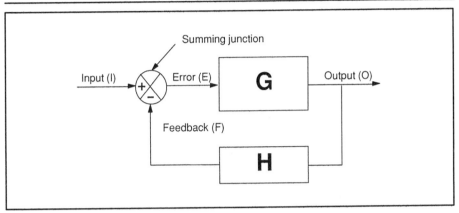

Figure 6.4

This system has three basic components

1. The SUMMING JUNCTION – which adds the negative feedback signal to the positive input signal. The signal which results from the summing junction is called the ERROR signal (E).

2. The FORWARD TRANSFER FUNCTION (G) – which represents all of the components of the system between the input and the load, such as the amplifier, servo valve, and hydraulic actuator.

3. The FEEDBACK TRANSFER FUNCTION (H) – which is a sensor of some kind. It measures the system output (position, velocity or force) and provides a compatible signal back to the summing junction.

Now we have a system that is a bit more complicated than the straightforward relationship shown in Figure 6.1. The input is multiplied by "G" and fed to the output. The output is multiplied by "H" and fed back to the input.

The signals are running around in a circle! Fortunately, we can simplify our view of this CLOSED LOOP system, by using it's OPEN LOOP GAIN.

Open Loop Gain is an important factor in determining the accuracy and response of a closed loop system.

Open Loop Gain

The OPEN LOOP GAIN, also referred to as K_V, is an important property of a closed loop system.

To illustrate the calculation of K_V, we'll use a typical position control system, as shown in Figure 6.5:

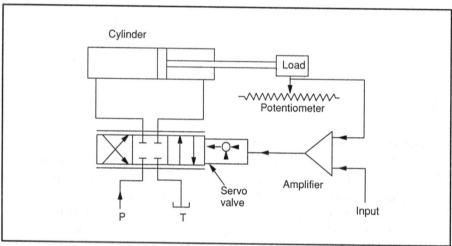

Figure 6.5

This system can be represented in block form (Figure 6.6):

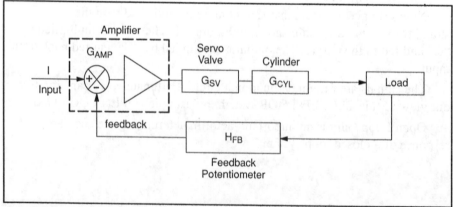

Figure 6.6

Let's consider each transfer function individually:

G_{AMP} – AMPLIFIER GAIN

The amplifier produces an OUTPUT CURRENT that is proportional to INPUT VOLTAGE. The amplifier's gain is expressed in terms of Milliamps of output per Volt of input (ma/volt).

G_{SV} – SERVO VALVE GAIN

The valve creates an OUTPUT FLOW RATE that is proportional to INPUT CURRENT from the amplifier. The valve's gain is expressed as in^3/sec of output flow per milliamp of current input. (in^3/sec/ma or in^3/sec ma).

G_{CYL} – CYLINDER GAIN

The cylinder converts an INPUT FLOW RATE of hydraulic fluid into a mechanical SPEED of motion OUTPUT. Its unit of gain is in/sec of output speed per in^3/sec of input flow. This gain happens to be the same as 1/(cylinder area) = 1/A (in/sec/in^3/sec or 1/in^2).

H_{FB} – FEEDBACK GAIN

The feedback transducer converts an input mechanical motion (inches) into an output voltage. (volts/inch).

The open loop gain can now be easily calculated by multiplying all of the FORWARD TRANSFER FUNCTIONS and the FEEDBACK TRANSFER FUNCTION together:

$$K_V = (G_{AMP})(G_{SV})(G_{CYL})H_{FB}$$

Equation #19

Make sure that the units that you use are consistent – do not mix metric with U.S. units or minutes with seconds, etc.

Example 6.1:

Calculate the open loop gain of the system shown in Figure 6.7:

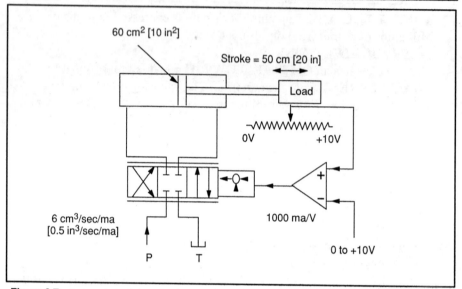

Figure 6.7

G_{AMP} = 1000 ma/volt

G_{SV} = 0.5 in³/sec/ma

G_{CYL} = 1/A = 1/10 = .1 in⁻²

H_{FB} = 10v/20 in = .5 volts/in

By equation #19:

$$K_V = (1000 \ ma/volt) \ (.5 \ in^3/sec \ ma) \ (.1/in^2) \ (.5 \ volts/in)$$

$$K_V = 1000 \times .5 \times .1 \times .5 \times \frac{ma \ in^3 \ volts}{volts \ sec \ ma \ in^2 \ in} = 25/sec$$

In order to calculate the open loop gain as shown, it is necessary to know the gains of each individual component. Sometimes, this gain information is not readily available. In addition, amplifier gain is usually adjustable, and therefore is not a constant.

As we'll see later, the open loop gain can also be estimated by other system factors, and then used to find the gains of individual components. Ideally, we'd like the open loop gain to be very large. But in practice, other characteristics of the system will act to limit its maximum value.

Open Loop Gain is also referred to as the VELOCITY CONSTANT or VELOCITY ERROR CONSTANT, since it affects the system's speed of response.

System Response

A typical closed loop position control system, like the one shown in Figure 6.6, might react to a step change in input as shown below.

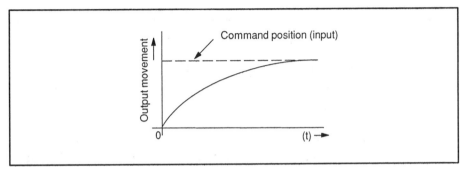

Figure 6.8

The output curve is not a straight-line response, as you can see.

This response is referred to as an EXPONENTIAL curve.

The curve rises rapidly at the start (high velocity), but gradually levels off (reduced velocity).

This is caused by the fact that, when the motion begins the error signal is large, opening the valve wide.

But as the actuator approaches the commanded position, the error signal gets smaller and smaller, closing the valve more and more.

An important part of an EXPONENTIAL CURVE is a variable known as the TIME CONSTANT, indicated by the Greek letter τ (tau).

One TIME CONSTANT is the amount of time it would take for the curve to reach final value if it could continue at its initial rate (Figure 6.9):

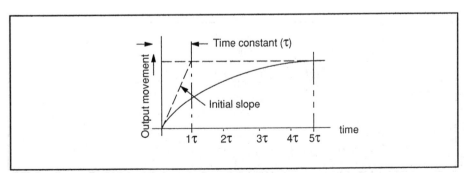

Figure 6.9

The initial slope of the curve is shown above.

But because the curve gradually levels off, the actual time to reach the maximum value (to within 1%) is actually equal to 5 TIME CONSTANTS.

The duration of one time constant can be calculated from the expression:

$$\tau = \frac{\text{Distance to be moved, } X_0 \ (in)}{\text{Initial velocity, } V_0 \ (in/sec)} = sec$$

Equation #20

How do we find X_0 and V_0?

The distance to be moved, X_0, can be found by taking the original input signal and dividing it by the feedback gain:

$$X_0 = \frac{I \ (volts)}{H_{FB} \ (volts/in)} = in$$

Equation #21

The initial velocity, V_0, is:

$$V_0 = I(G_{AMP})(G_{SV})(G_{CYL})$$

Equation #22

Substituting equations 21 and 22 into equation 20, we get:

$$\tau = \frac{X_0}{V_0} = \frac{\frac{I}{H_{FB}}}{I(G_{AMP})(G_{SV})(G_{CYL})}$$

Therefore,

$$\tau = \frac{1}{(G_{AMP})(G_{SV})(G_{CYL})H_{FB}}$$

Equation #23

And since, from equation #19,

$$(G_{AMP})(G_{SV})(G_{CYL})(H_{FB}) \;=\; K_V$$

We can simply state that:

$$\tau \;=\; \frac{1}{K_V}$$

Equation #24

This equation directly proves that there is a relationship between the response of the system (τ) and the open loop gain of the system (K_V).

If the gain is increased, the time constant is reduced, producing faster response.

Example 6.2:

For the system shown in Figure 6.7, calculate how long it will take this system to respond to a .2 volt step input. Assume that maximum valve drive current from the amplifier is 250 ma and that the amplifier's gain G_{AMP} = 1000 ma/volt.

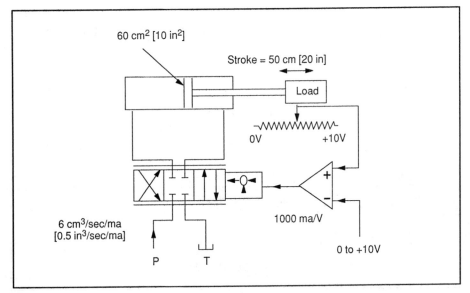

Figure 6.7

Since the input is a .2 volt step, and the amplifier gain is 1000 ma/volt, the amplifier's output will be:

$$.2 \ volts \times 1000 \ ma/volt \ = \ 200 \ ma$$

This is less than the maximum amplifier output of 250 ma, so the amplifier will not be saturated and the output can be considered an exponential curve.

From example 6.1 we already know that:

$$K_V \ = \ 25/sec$$

Therefore,

$$\tau \ = \ \frac{1}{K_V} \ = \ \frac{1}{25} \ = \ .04 \ sec$$

The time to come within 1% of final position is 5τ, so:

$$5\tau \ = \ 5 \times .04 \ sec \ = \ .20 \ sec$$

If faster response is needed, then the value of the open loop gain must be increased, resulting in a lower time constant. This is usually achieved by increasing the gain of the amplifier.

As the cylinder velocity increases, the effects of the load's inertia become more and more significant. Increasing the gain beyond a certain point will cause the cylinder to OVERSHOOT the commanded position.

This may be followed by a series of gradually decreasing undershoots and overshoots until the system eventually settles at the commanded position (Figure 6.10):

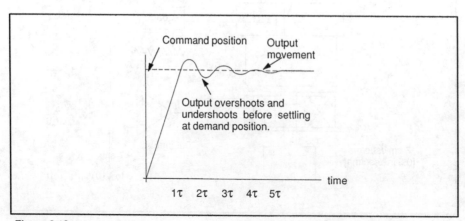

Figure 6.10

In some cases, a gain that is too high can cause the system to go UNSTABLE. This is a situation where the undershoots and overshoots do not die out, but get worse, resulting in loss of control of the load (Figure 6.11):

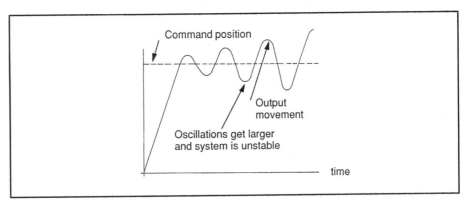

Figure 6.11

The maximum practical gain value that ensures system stability and acceptable settling time is determined by:

M – THE LOAD'S MASS

The larger the mass we're moving, the greater its inertia when it is moving. Since it is hard to stop a large moving mass, the system will have more of a tendency to oscillate.

C – THE STIFFNESS OF THE ACTUATOR

Whether hydraulic or mechanical, a system with low stiffness will also have a greater tendency to oscillate.

ξ – THE DAMPING COEFFICIENT

This is a measure of how much energy the system can dissipate to reduce oscillations. Natural damping in the system tends to reduce oscillations and limit loop gain.

F_V – VALVE FREQUENCY RESPONSE

The control valve itself takes a finite amount of time to respond. This response time must be taken into account.

The frequency response of the feedback transducer is usually not a limiting factor, unless it is installed in a very fast-response system. If the transducer's natural frequency is less than ten times the system's natural frequency, the transducer may also have to be taken into account.

Damping factors are very difficult to predict accurately without actual measurement, and usually change significantly over time. Mechanical friction and hydraulic internal leakage (across piston seals, valve spools, etc) are the main contributors to the damping coefficient. The damping coefficient in a hydraulic system is normally between 0.05 and .3, with the value of 0.2 being a good trial value for use in initial calculations.

The load's Mass (M) and the actuator hydraulic stiffness(C_H) are usually combined into one convenient factor, referred to as the LOAD NATURAL FREQUENCY, or ω_L .

The relationship between M, C_H and ω_L is:

$$\omega_L = \sqrt{\frac{C_H}{M}}$$

Equation #25

The actuator hydraulic stiffness may have to take into account the mechanical stiffness of the mounting. For example, the presence of rubber mounts or shock absorbing devices will tend to decrease actuator stiffness.

We are concerning ourselves with the natural frequency of the load, the valve and the feedback transducer in order to establish the natural frequency of THE SYSTEM (ω_S). We will establish how to do this later in the text.

However, we eventually need to know ω_s because we can use it to ensure that our system will not become unstable. In order to avoid producing an unstable system, gains must be used so that:

$$K_V \leq 2\xi\omega_S$$

Equation #26

Choosing a K_V of less than $2\xi\omega_S$ will ensure that the system does not become unstable. However, a very conservative gain setting may result in a long settling time.

To ensure acceptable settling time while ensuring stability, the maximum value of gain should be set to:

$$K_V(max) = \xi\omega_S$$

Equation #27

Before we go any further into the concept of "LOAD NATURAL FREQUENCY," we need to get a firm grasp on the concept of "hydraulic stiffness."

We will return to ω_L later.

Hydraulic Stiffness

In classical hydraulic systems, the hydraulic fluid was regarded as being INCOMPRESSIBLE. This is not absolutely correct.

When placed under pressure, hydraulic fluid (or any other substance) DOES compress, in much the same way as a SPRING. (Figure 6.12):

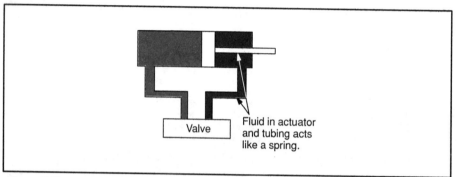

Valve

Fluid in actuator and tubing acts like a spring.

Figure 6.12

For slow moving or lightly loaded systems, the amount of compression is so slight that it can be ignored. However, in fast acting servo systems with high dynamic loads, IT CANNOT BE IGNORED any longer.

In fact, the stiffness of the hydraulic fluid may become the limiting factor in the overall performance of the system.

To maximize system performance, the stiffness of the fluid should be as high as possible.

When loads are rapidly accelerated and decelerated, high pressures can be created within the fluid. If the stiffness is too low, a considerable amount of compression can be created. When this compression is eventually released, it can cause oscillation of the load and lengthen settling times.

Factors that affect stiffness of an actuator are basically its size, shape and type of fluid used. This is illustrated by Figure 6.13:

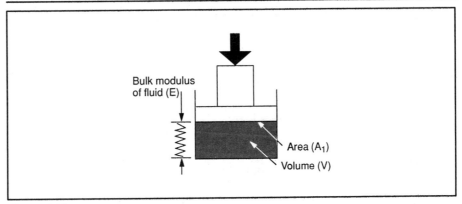

Figure 6.13

The stiffness of this actuator is determined by:

- The area of the piston (A, in^2)
- The volume of fluid under compression (V, in^3)
- The bulk modulus of elasticity of the fluid (E, lb$_f$/in^2)

The bulk modulus, E, is a measure of how easily the fluid can be compressed. The lower the E-value, the more compressible the fluid is.

A typical value of the bulk modulus of hydraulic oil is:

$$E = 2 \times 10^5 \; lb_f/in^2$$

Equation #28

The stiffness (sometimes called "spring rate") of the actuator can be calculated from the formula:

$$C_H = \frac{E(A_1)^2}{V}$$

Equation #29

Figure 6.14 illustrates the concept of stiffness in a typical servo valve and cylinder system.

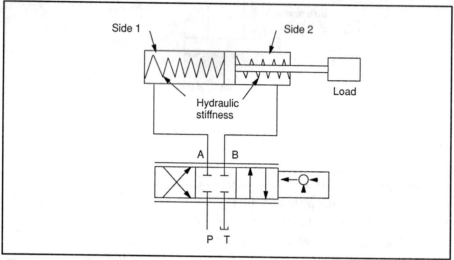

Figure 6.14

Whether the valve used is a servo valve or a proportional valve, flow will normally be metered in and out of the cylinder in both directions of movement. Both the P to A and B to T flow paths are restricted, and both sides of the piston will be pressurized during movement of the cylinder.

This is mechanically represented by Figure 6.15:

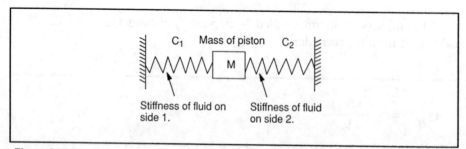

Figure 6.15

"Stiffnesses" can act either in series or in parallel. The simplest way to view the concept of stiffness is to look at it as a spring. To find the total effect of two or more springs, they are combined in the same way as capacitors in an electrical circuit (Figure 6.16):

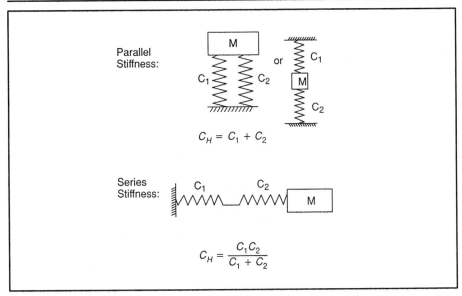

Figure 6.16

In the case of a hydraulic cylinder, the stiffnesses of the two sides of the cylinder act in parallel. Therefore, the total stiffness (C_H) of the cylinder in Figure 6.14 is given by:

$$C_H = C_1 + C_2$$

Equation #30

C_H in a linear actuator is expressed in units of lb_f/in.

In order to consider the stiffness of the system, it is necessary to include not only the volume of fluid in the actuator, but also the fluid in the lines between the cylinder and the valve, since the fluid in the lines is also under pressure.

As shown in Figure 6.17, the volume of fluid in the hydraulic lines between the valve and the cylinder are referred to as V_{L1} and V_{L2}.

V_1 and V_2 represent the volumes of fluid in the cylinder itself.

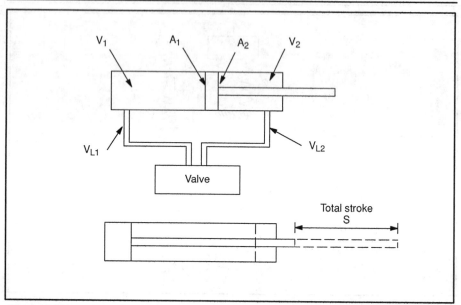

Figure 6.17

While V_{L1} and V_{L2} will remain almost constant, V_1 and V_2 will change constantly as the piston moves along its stroke.

If S represents the total stroke of the cylinder, and X_0 is the initial stroke at a given time, then for any initial stroke:

$$V_1 \; = \; A_1(X_0)$$

$$V_2 \; = \; A_2(S - X_0)$$

Equation #31

Using equations 29 and 30, we now have enough information to figure out the TOTAL HYDRAULIC STIFFNESS FOR AN UNEQUAL AREA ACTUATOR.

Total Volume (V) on Side 1 is $V = V_{L1} + V_1$

Total Volume (V) on Side 2 is $V = V_{L2} + V_2$

Side 1 Stiffness is $C_1 \; = \; \dfrac{E(A_1)^2}{V_{L1} + V_1}$ (Equation 29)

Side 2 Stiffness is $C_2 \; = \; \dfrac{E(A_2)^2}{V_{L2} + V_2}$ (Equation 29)

Total hydraulic stiffness is $C_H = C_1 + C_2$ (Equation 30)

$$C_H = \frac{E(A_1)^2}{V_{L1} + V_1} + \frac{E(A_2)^2}{V_{L2} + V_2}$$

OR

$$C_H = E\left[\frac{(A_1)^2}{V_{L1} + V_1} + \frac{(A_2)^2}{V_{L2} + V_2}\right]$$

Equation #32

Where: $C_H = lb_f/in$
$E = lb_f/in^2$
A_1 and $A_2 = in^2$
V_{L1}, V_{L2}, V_1, and $V_2 = in^3$

As the piston moves, V_1 and V_2 change constantly, causing the stiffness to change constantly. Figure 6.18 shows the stiffness variation for an unequal area cylinder.

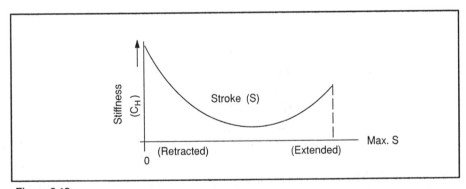

Figure 6.18

As can be seen, the stiffness is highest at either end of the stroke.

Stiffness reaches a MINIMUM somewhere around mid-stroke. For our performance calculations, it is this worst-case stiffness we are looking for, since it will have the greatest effect on our system performance.

The position where MINIMUM STIFFNESS occurs is referred to as X_m, and it can be calculated by the following equation:

$$X_m = \frac{\sqrt{R}\left(\frac{V_{L2}}{A_2} + s\right) - \frac{V_{L1}}{A_1}}{1 + \sqrt{R}}$$

Equation #33

Where: X_m = stroke length for minimum stiffness (in)

  R = area ratio = A_1/A_2

If the cylinder area ratio (R) is not large, and the line volumes (V_{L1} & V_{L2}) are relatively small when compared to the cylinder volumes, then it is sufficient to assume that minimum stiffness will occur at MID-STROKE. In other words:

$$X_m = \frac{S}{2}$$

Example 6.3:

Calculate the MINIMUM stiffness of the actuator in Figure 6.19:

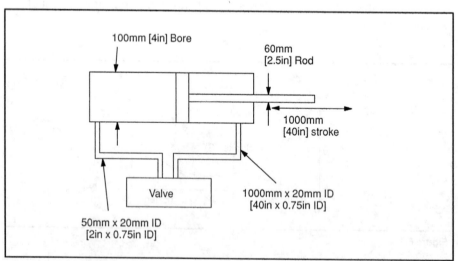

Figure 6.19

$$C_H\,(minimum) = E\left[\frac{(A_1)^2}{V_{L1} + V_1} + \frac{(A_2)^2}{V_{L2} + V_2}\right] \text{ where } E = 200,000\ lbf/in^2$$

$$A_1 = \frac{\pi(diameter)^2}{4} = \frac{3.1417(4)^2}{4} = 12.566\ in^2$$

$$A_2 \;=\; A_1 - A_{rod} \;=\; 12.566 - \frac{\pi(2.5)^2}{4} \;=\; 12.566 - 4.909 \;=\; 7.657 \; in^2$$

$$V_{L1} \;=\; area \times length \;=\; \frac{\pi(.75)^2}{4} \times 2 \;=\; .883 \; in^3$$

$$V_{L2} \;=\; area \times length \;=\; \frac{\pi(.75)^2}{4} \times 40 \;=\; 17.671 \; in^3$$

$$V_1 \;=\; \frac{S}{2} \times A_1 \;=\; \frac{40}{2} \times 12.566 \;=\; 251.32 \; in^3$$

$$V_2 \;=\; \frac{S}{2} \times A_2 \;=\; \frac{40}{2} \times 7.657 \;=\; 153.14 \; in^3$$

$$C_H \, (minimum) \;=\; 200,000 \left[\frac{12.566^2}{(.883 + 251.32)} + \frac{7.657^2}{(17.671 + 153.14)} \right]$$

$$C_H \, (minimum) \;=\; 193,868.6 \; lb_{f/in}$$

In the case of a DOUBLE ROD, or equal area cylinder, the analysis gets even easier.

In an equal area actuator, A_1 and A_2 are equal. That means the two volumes, at mid-stroke, are equal.

Therefore, at mid-stroke, the STIFFNESSES of the two sides must also be equal.

Stiffness of a double-rod (equal area) cylinder

Stiffness on side 1:

$$C_1 = \frac{E(A_1)^2}{V_{L1} + V_1}$$

Stiffness on side 2:

$$C_2 = \frac{E(A_2)^2}{V_{L2} + V_2}$$

If $A_1 = A_2 = A$, then $V_1 = V_2 = V$

Also assume $V_{L1} = V_{L2} = V_L$

Then,

$$C_1 = \frac{E(A)^2}{V_L + V}$$

$$=$$

$$C_2 = \frac{E(A)^2}{V_L + V}$$

<u>AND</u>

$$C_H = C_1 + C_2 = 2\frac{E A^2}{V_L + V}$$

Equation #34

If we further assume that the line volumes to this cylinder are equal, you can see that the total minimum stiffness is merely twice the minimum stiffness on one side.

We can use a similar analysis for hydraulic motor applications. A motor can also be viewed as an equal area actuator, because a certain flow produces a certain speed of revolution, regardless of direction of rotation.

However, a motor is a rotary device – not a linear actuator. This produces some differences in the way we measure "area," "volume" and stiffness.

Figure 6.20 illustrates a servo valve controlled hydraulic motor application.

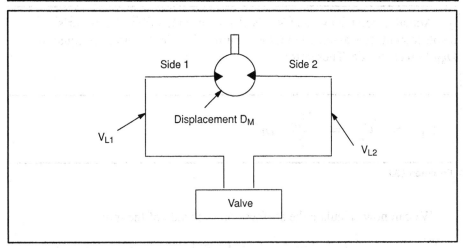

Figure 6.20

The basic formula for determining stiffness still applies to motors:

$$C_H = \frac{E(A)^2}{V}$$

However, the "area" and "volume" of a motor now need to be defined in order to proceed.

For a motor, the effective area (A_M) can be determined from the displacement of the motor (D_M).

D_M is expressed in in^3/revolution. For calculation purposes, we must convert "revolutions" to "radians," where:

1 revolution = 2π radians

The effective area of a motor is given by the expression:

$$A_M = \frac{D_M}{2\pi} \frac{in^3}{radian}$$

Equation #35

Notice that the "<u>effective</u>" area of a motor is expressed in cubic inches, not in square inches as you would expect for an area measure.

Another difference between cylinder and motor applications is that for motors, the volume of fluid in the actuator (the motor) is relatively small compared to the fluid that can be in the lines. Line volume will now constitute the majority of V, the total volume of fluid.

Actuator volume (V_1 and V_2) is also determined from the motor's displacement. It is assumed that the volume of oil in the motor is equal to $D_M/2$ on each side. Therefore:

$$V_1 = V_2 = \frac{D_M}{2} \; in^3$$

Equation #36

We can now calculate the stiffness on each side of the motor.

$$C_H = C_1 + C_2$$

$$C_1 = E\left(\frac{D_M}{2\pi}\right)^2 \times \left[\frac{1}{V_{L1} + \frac{D_M}{2}}\right]$$

$$C_2 = E\left(\frac{D_M}{2\pi}\right)^2 \times \left[\frac{1}{V_{L2} + \frac{D_M}{2}}\right]$$

Total stiffness for a hydraulic motor is therefore calculated using the expression:

$$C_H = E\left(\frac{D_M}{2\pi}\right)^2\left[\frac{1}{V_{L1} + \frac{D_M}{2}} + \frac{1}{V_{L2} + \frac{D_M}{2}}\right]$$

Equation #37

C_H, for a rotary actuator, is measured in (lb$_f$ in)/radian, which is an amount of TORQUE per unit of REVOLUTION, as opposed to C_H for a LINEAR actuator, which is measured in lb$_f$/in, a measure of FORCE per unit of LINEAR DISTANCE.

The concept of MINIMUM STIFFNESS does not apply to motors, since the volume of fluid in the motor does not change significantly as the motor rotates.

Therefore, the stiffness remains constant during operation:

$$C_H(min) = C_H$$

Equation #38

Example 6.4:

Calculate the stiffness of the rotary actuator shown in Figure 6.21:

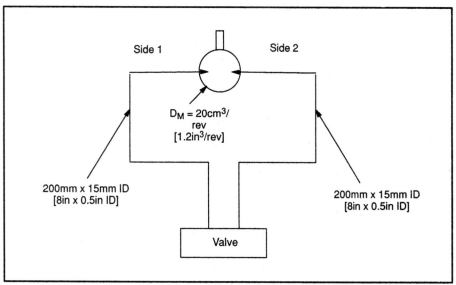

Figure 6.21

$$E = 200,000 \; lb_f/in^2$$

$$V_{L1} = V_{L2} = \frac{\pi \, d^2}{4} \times length = \frac{\pi(.5 \, in)^2}{4} \times 8 \, in = 1.57 \, in^3$$

$$C_H = 200,000 \times \left(\frac{1.2}{2\pi}\right)^2 \left[\frac{1}{1.57 + \frac{1.2}{2}} + \frac{1}{1.57 + \frac{1.2}{2}} \right]$$

$$C_H = 6,723.6 \frac{lb_f \, in}{radian}$$

Since the volume of fluid under compression is an important factor in determining stiffness, servo type actuators (with servo valve built onto the actuator) have an inherent advantage over standard cylinders and motors used with a separate servo valve. In a servo type actuator, line volume is held to a minimum, resulting in a stiffer system. <u>Minimum line volume gives maximum stiffness.</u>

The mechanical stiffness of any linkages or mechanisms between the actuator and the load must be considered, as well as any lack of rigidity in the actuator mounting itself.

Mechanical stiffness will usually act in series with the hydraulic stiffness (Figure 6.22):

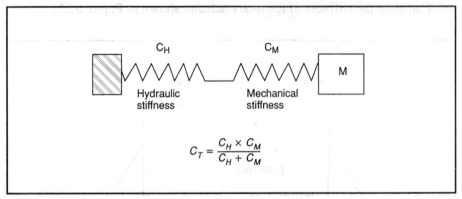

Figure 6.22

C_T = total stiffness of actuator and mechanical connections.

In most well designed systems, the mechanical stiffness should be much greater than the hydraulic stiffness.

If mechanical stiffness is high enough, indicating rigid mounting of the actuator and lack of "flexing" in any linkages, then the mechanical stiffness can be ignored.

However, there can be a surprisingly low mechanical stiffness in complex linkages or fixtures. In such cases, an analysis of the mechanical stiffness is a wise idea.

Load Natural Frequency

Now that you've learned what "stiffness" is, and how to find it, you can calculate it for a particular load/actuator combination and then use it to find the NATURAL FREQUENCY of our load/actuator assembly.

NATURAL FREQUENCY is an important concept, because it gives us a measure of the conditions under which load control can be lost, due to instability.

Figure 6.23 provides an example of the NATURAL FREQUENCY of a system.

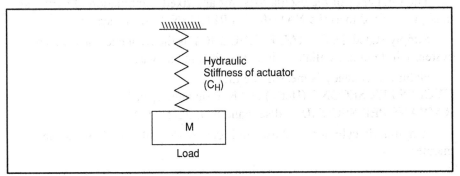

Figure 6.23

This simple system consists of a load hung from a spring. If the load is pulled down a short distance and released, the load will obviously bounce up and down as shown in Figure 6.24:

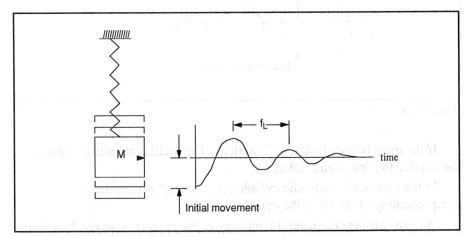

Figure 6.24

The bouncing action is referred to as OSCILLATION.

If you could attach a pen to the mass and pull a sheet of graph paper horizontally behind the mass so that the pen could mark the paper, you would get a trace of the oscillations as shown in Figure 6.24.

The oscillations get smaller and eventually die out due to air resistance and friction in the spring.

How fast the system will oscillate depends on the stiffness of the spring and the amount of mass hung on the spring. The system will oscillate more rapidly if we use a stiffer spring, and will oscillate more slowly if we increase the mass hung on the spring.

The frequency of the oscillations, for any fixed combination of spring and mass, is referred to as the NATURAL FREQUENCY of the system.

Simply stated, NATURAL FREQUENCY is the frequency at which a system will tend to oscillate, if it is disturbed in any way.

Natural Frequency is measured either in
CYCLES PER SECOND (Hertz) and is designated f_L or in
RADIANS PER SECOND (rad/sec) and is designated ω_L .

A hydraulic cylinder, as shown in Figure 6.25, will react in the same manner.

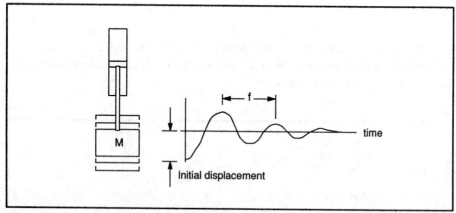

Figure 6.25

If the mass is disturbed, it will oscillate. The oscillations will eventually die out due to friction and leakage.

In the case of the hydraulic cylinder, the "spring" is provided by the compressibility of the oil in the cylinder.

We can calculate the natural frequency of the cylinder/load combination from the following formula:

$$\omega_L = \sqrt{\frac{C_H}{M}}$$

Equation #39

Where: ω_L = natural frequency (radians/second)
 C_H = hydraulic stiffness (lbf/inch)
 M = mass of the load (lbf. sec^2/inch)

A handy conversion factor for converting Hz to radians/second, or vice versa, is:

Hz to rad/sec: $f = \omega/2$
rad/sec to Hz: $\omega = 2f$

Using the Hz to rad/sec conversion factor, we can therefore convert equation #39 to Hz by:

$$f_L = \frac{1}{2\pi} \sqrt{\frac{C_H}{M}}$$

Equation #40

Where: f_L = natural frequency (Hz)

The natural frequency of a motor is calculated in much the same way. However, the motor is a ROTARY device – unlike the linear cylinder. We also speak of TORQUE in motor applications, as opposed to the FORCE we generate with a cylinder.

Therefore, the natural frequency of a motor is calculated from:

$$\omega_L = \sqrt{\frac{C_H}{J}}$$

Equation #41

Where: C_H = hydraulic motor stiffness (lb$_f$ in/rad)
 J = load INERTIA (in lb$_f$ sec^2)

Using our conversion factor for rad/sec to Hz, it follows that:

$$f_L = \frac{1}{2\pi} \sqrt{\frac{C_H}{J}}$$

Equation #42

It is very important to realize that equations 39 through 42 apply to the mass being moved AS SEEN BY THE ACTUATOR. If the actuator is directly connected to the load, the calculation is easy. However, if the actuator is connected to levers, pulleys or gears that actually move the load, these devices will probably serve to either increase or decrease the MECHANICAL ADVANTAGE of the actuator.

Figure 6.26 shows just a few possible load configurations, where the actuator is NOT directly connected to the load mass.

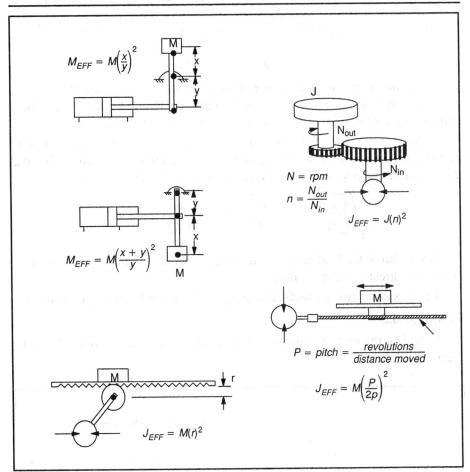

$$M_{EFF} = M\left(\frac{x}{y}\right)^2$$

$$M_{EFF} = M\left(\frac{x + y}{y}\right)^2$$

$$N = rpm$$

$$n = \frac{N_{out}}{N_{in}}$$

$$J_{EFF} = J(n)^2$$

$$P = pitch = \frac{revolutions}{distance\ moved}$$

$$J_{EFF} = M\left(\frac{P}{2p}\right)^2$$

$$J_{EFF} = M(r)^2$$

Figure 6.26

The presence of these various linkages between actuator and load will change the force that the actuator has to actually exert to move the load weight.

If you analyzed the forces involved, using some basic physics, you would eventually find that these various linkages cause the actuator to see a different value of MASS than the load actually has.

This different value of mass is referred to as the EFFECTIVE MASS.

$$M_{EFF}$$

The first example in Figure 6.26 shows a cylinder moving a mass via a pivoting lever. If the distance from the pivot to the load ("x") is 10 inches and the distance from the pivot to the cylinder rod ("y") is 40 inches, and the mass of the load is 100 lb_m, then the effective mass seen by the cylinder is:

$$M_{EFF} = 100 \; lb_m \times \left(\frac{10}{40}\right)^2 = 100 \; lb_m \times \left(\frac{1}{4}\right)^2 = 100 \; lb_m \times \frac{1}{16}$$

$$M_{EFF} = 6.25 \; lb_m$$

Using the mechanical advantage of the lever, we have effectively decreased the mass to be moved by the cylinder to 1/16th of what it was. The cylinder must exert only 6.25 lb_m of force to move the 100 lb_m load.

Further, if we assume that the stiffness of the lever is very high (i.e. much higher than the hydraulic stiffness), then from equation #39 we can predict that the natural frequency of the system will be 4 times higher than before:

$$\sqrt{\frac{C_H}{\frac{1}{16}M}} = \sqrt{\frac{16C_H}{M}} = 4\sqrt{\frac{C_H}{M}} = 4\omega_L$$

If we disturb the load mass, it will oscillate 4 times faster than if it were connected directly to the cylinder.

Let's examine two worked-out examples of natural frequency calculation.

Example 6.5:

Calculate the natural frequency of the load/actuator combination shown in Figure 6.27:

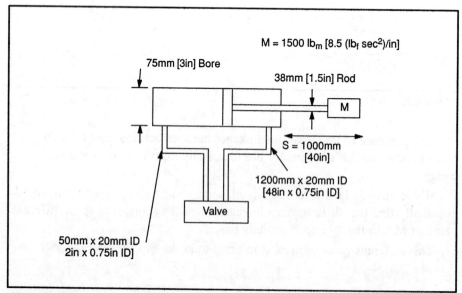

M = 1500 lb_m [8.5 (lb_f sec²)/in]

75mm [3in] Bore

38mm [1.5in] Rod

M

S = 1000mm [40in]

1200mm x 20mm ID [48in x 0.75in ID]

Valve

50mm x 20mm ID 2in x 0.75in ID]

Figure 6.27

$$A_1 = \frac{\pi(d)^2}{4} = \frac{3.1417 \times (3 \; in)^2}{4} = 7.07 \; in^2$$

$$A_2 = \frac{\pi(d)^2}{4} = A_1 - \frac{3.1417 \times (1.5\ in)^2}{4} = 5.30\ in^2$$

$$V_{L1} = \frac{\pi(d)^2}{4} \times L = \frac{3.1417 \times (.75\ in)^2}{4} \times 2\ in = .884\ in^3$$

$$V_{L2} = \frac{\pi(d)^2}{4} \times L = \frac{3.1417 \times (.75\ in)^2}{4} \times 48\ in = 21.206\ in^3$$

$$X_m = \frac{S}{2} = \frac{40\ in}{2} = 20\ in$$

$$C_H = 200,000\left(\frac{7.07^2}{.884 + 20(7.07)} + \frac{5.30^2}{21.206 + 20(5.30)}\right) = 114,425.3$$

$$\omega_L = \sqrt{\frac{C_H}{M}} = \sqrt{\frac{114,425.3}{8.5}} = 116\ rad/sec$$

$$f_L = \frac{\omega_L}{2\pi} = \frac{116}{2 \times 3.1417} = 18.5 Hz$$

Example 6.6:

Calculate the natural frequency of the motor drive system shown in Figure 6.28:

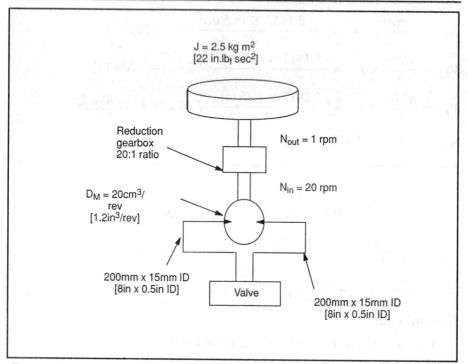

Figure 6.28

The displacement of the motor and the line volumes are the same as example 6.4, therefore:

$$C_H = 6,723.6 \frac{lb_f \ in}{radian}$$

Load Inertia is given in the problem:

$$J = 22 \ in \ lb_f \ sec^2$$

However, there is a reduction gearbox between the motor and the load. The equation for this situation is given in Figure 6.26 as:

$$J_{EFF} = J(n)^2 = 22\left(\frac{1}{20}\right)^2 = .055 \ in \ lb_f \ sec^2$$

We now use the EFFECTIVE inertia and equation #41 to form a new equation:

$$\omega_L = \sqrt{\frac{C_H}{J_{EFF}}}$$

Equation #43

$$\omega_L = \sqrt{\frac{6,723.6}{.055}} = 350\,rad/sec$$

Using the conversion factor from radians to Hertz once again,

$$f_L = \frac{\omega_L}{2\pi}$$

Equation #44

$$f_L = \frac{350}{2 \times 3.1417} = 55.7\,Hz$$

The NATURAL FREQUENCY is important because it represents a situation which is to be avoided when designing a system.

If the system tends to OSCILLATE when disturbed, <u>it becomes important not to disturb the system at a rate which will feed these oscillations and make them larger</u>.

The rate at which the system is INTENTIONALLY DISTURBED (by moving the load) is determined by the OPEN LOOP GAIN of the system.

The higher the gain, the faster the attempt to move the load.

Therefore, the system's natural frequency will largely dictate how fast the load can be moved without losing control of it, and thus will determine the maximum gain that can safely be applied.

Determination of Open Loop Gain

As previously stated in equation #27, the maximum usable value of open loop gain (to ensure system stability and an acceptable settling time) can be determined from the expression:

$$K_V(max) = \xi \omega_S$$

In this case, ω_S represents the natural frequency of the COMPLETE CLOSED LOOP SYSTEM, which includes:

The Servo Valve
The Amplifier
The Actuator/Load Combination and
The Feedback Transducer

Each of these component pieces will in fact have its own NATURAL FREQUENCY. Generally, one of the closed loop components will have a natural frequency that is far lower than that of the other component (usually the actuator/load combination).

It is the component, or components, with the LOWEST NATURAL FREQUENCY that will dictate the maximum gain that can safely be used on the overall system.

In actual practice, the natural frequency of the amplifier and the feedback transducer are normally at least ten times higher (and sometimes hundreds of times higher) than that of the valve or actuator/load combination, and can be ignored.

The servo valve and the actuator/load natural frequencies are therefore focussed upon in determining the natural frequency of the system, ω_S.

Ideally, a servo valve should be chosen for an application where the natural frequency of the valve is three or more times the natural frequency of the actuator/load:

IDEAL SERVO VALVE NATURAL FREQUENCY IS $\omega_V > 3\omega_L$

where ω_V = natural frequency of valve
  ω_L = natural frequency of actuator/load

However, this may not always be achievable, particularly in pressure or force control applications involving larger cylinders and small oil volumes.

The natural (sometimes called "limiting") frequency of the valve can be obtained from the catalog data on the valve. The catalog data, however, typically shows the natural frequency under RATED OPERATING CONDITIONS only.

The natural frequency of the valve will vary somewhat at other values of system supply pressure, so the catalog will also typically provide a <u>correction</u>

<u>factor</u> that can be used to determine what the natural frequency of the valve will be under your particular operating conditions.

We therefore have three different situations that need to be considered before determining the open loop gain.

CASE A: If $\omega_V > 3\omega_L$
   (Valve's natural frequency is 3 or more times higher than that of the load/actuator), then the actuator/load combination is the limiting factor.

CASE B: If $3\omega_L > \omega_V > .3\omega_L$
   (Valve's natural frequency is between .3 and 3 times natural frequency of load/actuator), then both the valve and load/actuator limit the system.

CASE C: If $\omega_V < .3\omega_L$
   (Valve's natural frequency is less than .3 times the natural frequency of the load/actuator), then the servo valve is the limiting factor.

CASE A:

In such applications, the load/actuator combination is the limiting factor. It can be assumed that the natural frequency of the system (ω_S) is that of the load/actuator.

$$\omega_S = \omega_L$$
$$\text{and } K_V(max) = \xi \ (\omega_S)$$

The DAMPING COEFFICIENT, ξ, typically ranges between .05 and .3 in a hydraulic system, with an average value of .2. Therefore,

$$K_V(max) = .2(\omega_L) \ sec^{-1}$$

CASE B:

In this situation, the servo valve has to be considered along with the load/actuator arrangement. The combined natural frequency is found by using:

$$\omega_S = \frac{\omega_L \ (\omega_V)}{\omega_L + \omega_V}$$
$$\text{and } K_V(max) = \xi \ (\omega_S)$$

Therefore, for Case B, maximum system gain is determined by:

$$K_V(max) = .2\left(\frac{\omega_L \ (\omega_V)}{\omega_L + \omega_V}\right) sec^{-1}$$

CASE C:

Under these conditions, the valve itself is the limiting factor in the system. However, you can assume a higher damping factor of .4 in the open loop gain expression.

$$\omega_S = \omega_V$$

$$\text{and } K_V(max) = \xi\,(\omega_S)$$

Therefore, for Case C, maximum system gain is determined by:

$$K_V(max) = .4(\omega_V)\,sec^{-1}$$

Example 6.7:

Determine the maximum value of open loop gain for example 6.5, now shown below in Figure 6.29.

Assume that ω_V, the 90° phase shift point, is 40 Hz at the system's working pressure.

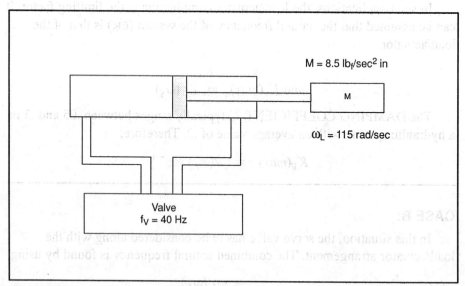

M = 8.5 lb$_f$/sec^2 in

M

ω_L = 115 rad/sec

Valve
f$_V$ = 40 Hz

Figure 6.29

Natural frequency of the load, from example 6.5, was:

$$\omega_L = 115 \; radians/second$$
$$\text{so } 3\omega_L = 345 \; radians/second$$
$$\text{and } .3\omega_L = 34.5 \; radians/second$$

Natural frequency of the valve, from Figure 6.29, is:

$$\omega_V \;=\; 2\pi\, f_V \;=\; 2\pi\,(40\,Hz) \;=\; 251\;radians/second$$

This is a CASE B situation, since ω_V is less than 3 x ω_L, but more than .3 ω_L. Therefore,

$$\omega_S \;=\; \frac{\omega_L\,(\omega_V)}{\omega_L + \omega_V} = \frac{115(251)}{115 + 251} \;=\; 78.9\;rad/sec$$

The natural frequency of this system is therefore 78.9 radians/sec, and the maximum allowable open loop gain is:

$$K_V(max) \;=\; .2(\omega_S) \;=\; .2(78.9) \;=\; 15.8\;sec^{-1}$$

Once the valve is sized, and the natural frequency and open loop gain of the system have been determined, it is now possible to make some estimates of how the system will perform.

As previously stated, certain approximations and simplifications have been made in this text, in order to keep the analysis as simple as possible.

Levels of performance predicted by using the methods in this manual must be regarded as conservative estimates, and not as an indication of the best possible performance that can be achieved.

Position Control System

A typical command signal, which could be sent to a closed loop system and cause a cylinder to move to a desired position, is shown in Figure 6.30:

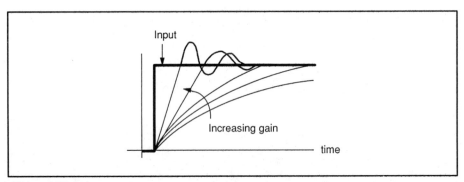

Figure 6.30

Five possible responses are shown for the given step input. As system gain is increased, the responses get faster. They take less time to rise to the commanded position.

Beyond a certain point, however, increasing the gain will cause the load to OVERSHOOT the commanded position. In other words, the load GOES BEYOND the intended stopping point because it is moving too fast to stop in time.

The open loop gain, K_V, is therefore a direct measurement of HOW FAST the system will move the load to the commanded position. If $K_V = 0.2\,\omega_S$, the load will reach the commanded position within 5 time constants, where 1 time constant is equal to $1/K_V$.

This is why K_V is sometimes referred to as the VELOCITY CONSTANT.

Open loop gain, in a position control system, is further described by referring to it as K_{VX}. The "X" in the subscript reveals that the gain being discussed is used in a POSITION CONTROL SYSTEM – not a velocity or force control system.

CALCULATING THE RESPONSE TIME OF A SYSTEM

K_{VX} can be used to predict the RESPONSE TIME of a position control system if its value is known, and if the system has been correctly set up, meaning:

All system components are working within rated capacity.

The pump is supplying the necessary flow.

The servo valve has been properly sized.

The amplifier's gain has been adjusted for maximum sensitivity to changes in the input signal.

The "PROPORTIONAL BAND" of the valve must also be known. Figure 6.31 will be used to illustrate this important principle.

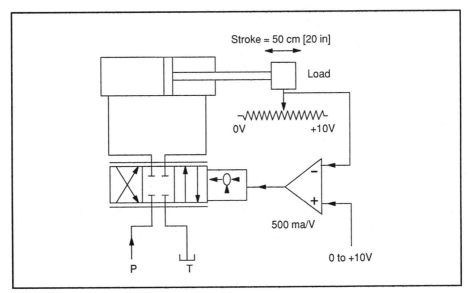

Figure 6.31

The gain of the amplifier is 500 ma of output per volt of input. In other words, the amplifier will attempt to output:

500 ma of current if 1 volt is applied to the input

1,000 ma of current if 2 volts is applied to the input

2,500 ma of current if 5 volts is applied to the input, etc.

Assume that the MAXIMUM RATED DRIVE CURRENT OF THE VALVE is 200 ma.

This means that the servo valve is WIDE OPEN when 200 ma is applied to the torque motor. When the valve is wide open, it is said to be **SATURATED**.

Sending more than 200 ma to the valve (driving it beyond the SATURATION POINT) will not accomplish anything, and may in fact burn out the valve's torque motor.

Therefore, most servo valve amplifiers have a CURRENT LIMITING adjustment, which sets the maximum current output of the amplifier. This adjustment prevents damage to the torque motor of the valve, by setting a maximum allowed output from the amplifier.

In the example above, assume further that the cylinder is fully retracted, and that a +8 volt step signal has been applied to the amplifier's input. The system in Figure 6.31 uses a command signal and a feedback signal of 0 to +10 volts

This implies that the commanded position is 16 inches, since +10 volts corresponds to 20 inches of movement.

$$\frac{20 \ inches}{10 \ volts} \times 8 \ volts \ = \ 16 \ inches$$

Since the command is now +8 volts, and the feedback is 0 volts, the error signal is +8 volts.

This would normally indicate an output of:

$$I \ = \ 8 \ volts \ (500 \ ma/volt) \ = \ 4,000 \ ma$$

However, the amplifier has been current limited to 200 ma of maximum output.

The valve and the amplifier are SATURATED, providing MAXIMUM FLOW to the cylinder and moving the load at MAXIMUM SPEED toward the commanded position.

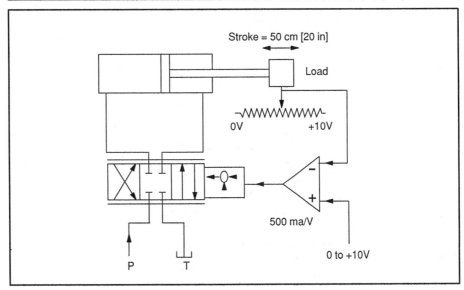

Figure 6.31

When will the load start slowing down?

The servo valve spool will only start to move back toward null, and reduce flow to the cylinder, when the amplifier output signal decreases from 200 ma.

This corresponds to an error signal of:

$$\frac{500 \; ma}{volt} \times \; ? \; error \;\; = \;\; 200 \; ma \; output$$

$$? \; error \;\; = \;\; \frac{200 \; ma}{500 \; ma/volt} \;\; = \;\; .4 \; volts \; error$$

If +10 volts represents 20 inches of movement, then .4 volts represents:

$$\frac{20 \; inches}{10 \; volts} \times .4 \; volts \;\; = \;\; .8 \; inch$$

To sum up the action, when an 8 volt step input is applied, the valve will fully open and move the load at maximum speed toward the 16 inch position.

When the load gets within .8 inch of commanded position (i.e. when it gets to 15.2 inches), the valve will begin to close and decelerate the load during the last .8 inch of motion.

In order to calculate how long this will take, it is necessary to calculate the amount of time it takes the load to travel from 0 inches to 15.2 inches at maximum velocity, then add to that the 5 time constants taken up by the remaining .8 inch of travel.

The last .8 inch of travel, where the valve and amplifier are operating below the saturation level, is referred to as the proportional band (Figure 6.32):

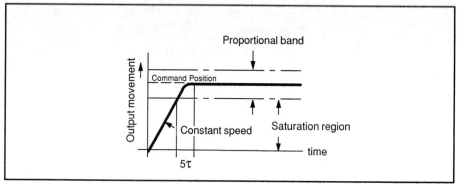

Figure 6.32

Once the proportional band is known, the response time of the system can be calculated.

Figure 6.33 provides the additional information required to calculate K_{VX}.

Example 6.8:

Calculate the response time to a step input of 5 volts for the system shown in Figure 6.33:

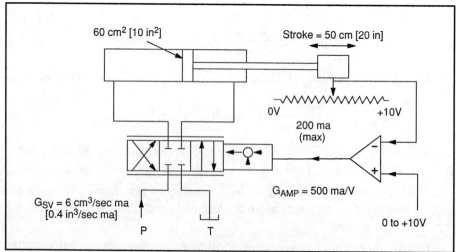

Figure 6.33

1. First, find the proportional band of the amplifier.

$$G_{AMP} = \frac{Output}{Input},$$

$$\text{therefore } Input = \frac{Output}{G_{AMP}} = \frac{200 \; ma}{500 \; ma/V} = .4 \; volts$$

If +10V input = 20 inches stroke, then .4V x 20 in/ 10 V = .8 in. Therefore, the proportional band = .8 inch.

2. Next, find the commanded position.

    If +10V input = 20 inches stroke, then +5V x 20 in/10 V = 10in

    Therefore, the commanded position is 10 inches.

3. The distance that the cylinder travels during SATURATION OF THE VALVE is therefore: 10 inches – .8 inch = 9.2 inches

4. In order to calculate the time required to travel 9.2 inches, the VELOCITY of the cylinder must be known. Cylinder velocity can be calculated from flow into the cylinder.

$$G_{SV} = \frac{OutputFlow}{InputCurrent} = \frac{Q_A}{I}$$

Therefore, $Q_A = G_{SV} \times I = .4 \frac{in^3/sec}{ma} \times 200 \; ma = 80 \; in^3/sec$

Cylinder velocity can now be calculated:

$$V_{max} = \frac{Q_A}{A_1} = \frac{80 \; in^3/sec}{10 \; in^2} = 8 \; in/sec$$

5. The time to travel 9.2 inches, at 8 inches per second, is:

$$T = \frac{9.2 \; in}{8 \; in/sec} = 1.15 \; seconds$$

6. The time to travel the remaining .8 inch is 5τ, where $\tau = 1/K_V$:

$$K_V = (G_{AMP})(G_{SV})(G_{CYL})H_{FB}$$

G_{SV} = .4 in^3/sec/ma
G_{AMP} = 500 ma/V
G_{CYL} = 1/10 in^2
H_{FB} = 10 V/20 in = .5 V/in

$$K_{VX} = \frac{.4 \times 500 \times .5}{10} = 10sec^{-1}$$

$$\tau = \frac{1}{K_{VX}} = \frac{1}{10 \; sec^{-1}} = 0.1 \; sec$$

The time to travel the remaining .8 inch is:

$$5\tau = 5 \times .1 \; sec = .5 \; second$$

7. The total response time for this motion is therefore:

$$T_{(total)} = T + 5\tau = 1.15 \; seconds + .5 \; second = 1.65 \; seconds$$

CALCULATION OF FOLLOWING ERROR

We have considered the response time of a system to an instantly changing step input. But what happens when the input is a RAMP signal?

As you recall, a ramp signal is a gradua lly changing signal used to reduce the shock of acceleration and deceleration in a system, when synchronizing the motion of two or more actuators, or to control the rate of application of forces. Figure 6.34 shows a simple ramp signal and a typical system response to it.

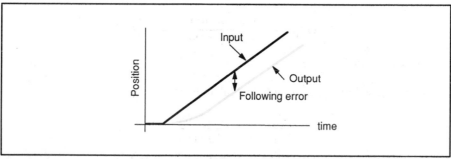

Figure 6.34

As the input is applied, there is a certain amount of time lag between the time that the input changes and the time that the load actually responds.

The amount of lag between the input signal and the output response is known as **FOLLOWING ERROR**.

Notice that the graph plots POSITION with respect to TIME. If the ramp signal is a sloping straight line then the cylinder is moving at a CONSTANT SPEED.

The higher the slope of the ramp, the faster the speed of the cylinder.

At any instant in time, after initial acceleration, the output is always a constant distance, or error, behind the input. For this reason, Following Error is also called "Velocity Misalignment."

The amount of following error is determined by the speed of motion, and by the gain of the system.

This is indicated by the following expression:

$$X_f = \frac{V}{K_{VX}}$$

Equation #45

Where: X_f = Following Error (inches)

V = Velocity (inches/second)

K_{VX} = Open Loop Gain (seconds^{-1})

The faster the input signal rises, the greater the velocity and the harder it will be for the cylinder to keep up with it. The faster the motion, the greater the following error.

However, if the GAIN is increased, the system will be sensitive to smaller changes in input. More gain will decrease the following error.

There are limits as to how much gain can be tolerated. As with the step input, increasing the gain too much can cause instability in the system, which has the effect shown in Figure 6.35:

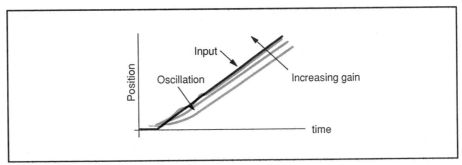

Figure 6.35

The same criteria used before applies for the prevention of instability.

$$K_V(max) = \xi \omega_S$$

Equation #46

CALCULATION OF POSITIONAL ERROR

The prime question in a position control system is "How accurately does the system control the position of the load?"

Many factors affect the accuracy that the system can achieve AND HOLD during operation. Both external and internal disturbances will tend to affect the static accuracy, and must be considered.

INTERNAL DISTURBANCES INCLUDE:

- Valve Hysteresis
- Valve Threshold
- Valve Null Shift (due to temperature/pressure variations)
- Amplifier Drift
- Accuracy of the Feedback Transducer

EXTERNAL DISTURBANCES INCLUDE:

- Reactive loads on actuators (Tool forces, Shock loads, etc.)
- Load Weight (Vertically mounted cylinders, etc.)

OTHER DISTURBANCES INCLUDE:

- Backlash in actuator mountings
- Feedback errors induced by the transducer.

Position errors caused by normal deviations in the valve and amplifier will be affected by:

- The size (rated flow) of the valve (Q_R)
- The open loop gain of the system (K_V)
- The actual or effective area of the actuator (A)

The following expression can be used to estimate the positional error caused by the valve and amplifier:

$$\Delta X_U = 0.04 \left(\frac{Q_{RP}}{K_{VX}(A)} \right)$$

Equation #47

Where: ΔX_U = Position error due to valve uncertainties (inches)
Q_{RP} = Rated valve flow at operating system pressure (in^3/sec)
K_{VX} = Open Loop Gain (sec^{-1})
A = Actuator area (in^2)

Additional errors, induced by the external forces, are affected by the magnitude of the external forces and by the system pressure. They can be estimated from the following expression:

$$\Delta X_E \;=\; 0.02 \left(\frac{Q_{RP}}{K_{VX}(A)} \right) \left(\frac{F_E}{P_S(A)} \right)$$

Equation #48

Where: ΔX_E = Position Errors caused by external forces (inches)

Q_{RP} = Rated valve flow at operating system pressure (in^3/sec)

K_{VX} = Open Loop Gain (sec^{-1})

A = Actuator area (in^2)

F_E = External Force (lb$_f$)

P_S = System pressure (psi)

Other induced errors which can occur are:

ΔX_{BL} = Backlash errors in actuator mounting or drive mechanisms.

ΔX_{FB} = Error induces by the feedback transducer.

Backlash errors are caused by play or looseness in the hydraulic actuator, and any play in devices connected between the hydraulic actuator and the load being moved.

Backlash acts to reduce the stiffness of the system, decreasing the natural frequency and making instability more possible.

Every effort must be made to eliminate backlash error through good design practice. Wear can also create or increase backlash in mechanical components, so proper maintenance is also important in high-performance machinery.

FEEDBACK ERROR is induced into the system by the feedback transducer. This error is typically found in the transducer manufacturers catalog or data sheet for the device. ALL feedback transducers will produce some error – but, generally speaking, the higher the quality of the transducer, the greater its accuracy.

THE TOTAL POSITION ACCURACY OF THE SYSTEM is now found by simply adding up the individual errors from each source:

$$\Delta X_{TOT} \;=\; \Delta X_U + \Delta X_E + \Delta X_{FB} + \Delta X_{BL}$$

Equation #49

Where: ΔX_{TOT} = Total Position Accuracy

The following is an example of an ACCURACY ESTIMATE for a POSITION control system.

Example 6.9:

Estimate the total position accuracy of the system shown in Figure 6.36. Assume that the cylinder is rigidly mounted, and that the feedback transducer is accurate to within .002 inch.

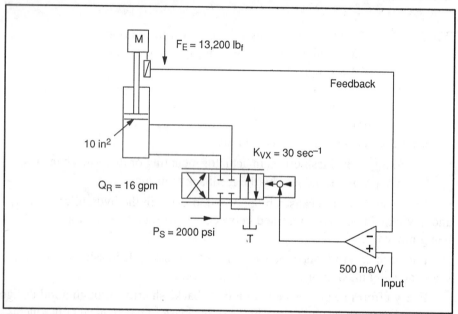

Figure 6.36

In this example, Q_R of the valve is 16 GPM. This means that the valve passes 16 GPM at 1000 PSI of pressure drop.

However, the system pressure is 2000 PSI – we need to know Q_{RP}, which is the rated flow at the given operating system pressure.

$$Q_{RP} = Q_R \sqrt{\frac{P_S}{1000}}$$

$$Q_{RP} = 16 \sqrt{\frac{2000}{1000}} = 22.6 \; gallons/minute$$

Here, Q_{RP} is calculated in GPM, but the formula for ΔX_U requires that Q_{RP} be stated in cubic inches per second, so:

$$Q_{RP} = 22.6 \frac{gallons}{minute} \times 231 \frac{in^3}{gallon} \times \frac{1 \; minute}{60 \; seconds} = 87.1 \; in^3/sec$$

From Figure 6.36, the problem statement and the calculation of Q_{RP},

$K_{VX} = 30/\text{sec}$
$Q_{RP} = 87.1 \text{ in}^3/\text{sec}$
$A_1 = 10 \text{ in}^2$
$F_E = 13,200 \text{ lb}_f$
$P_S = 2,000 \text{ PSI}$
$\Delta X_{FB} = .002 \text{ in}$
$\Delta X_{BL} = 0$

The position error due to the valve and amplifier can now be calculated:

$$\Delta X_U = 0.04 \left(\frac{Q_{RP}}{K_{VX}(A)} \right)$$

$$\Delta X_U = 0.04 \left(\frac{87.1}{30 \times 10} \right) = 0.04 \left(\frac{87.1}{300} \right) = 0.0116 \text{ inch}$$

The position error due to external forces can also be calculated:

$$\Delta X_E = 0.02 \left(\frac{Q_{RP}}{K_{VX}(A)} \right) \left(\frac{F_E}{P_S(A)} \right)$$

$$\Delta X_E = 0.02 \left(\frac{87}{30 \times 10} \right) \left(\frac{13,200}{2000 \times 10} \right) = 0.0038 \text{ inch}$$

Total position accuracy is therefore:

$$\Delta X_{TOT} = \Delta X_U + \Delta X_E + \Delta X_{FB} + \Delta X_{BL}$$

$$\Delta X_{TOT} = 0.0116 \text{ inch} + 0.0038 \text{ inch} + 0.002 \text{ inch} + 0$$

$$\Delta X_{TOT} = \pm 0.0174 \text{ inch}$$

The system will position the load accurately to within 17.4 thousandths of an inch of commanded position.

Velocity Control System

A Velocity Control System is very similar in appearance to a position control system.

However, as its name implies, its objective is to control the SPEED – not the position – of the actuator.

One of the primary differences between the Position Control and the Velocity Control is that the Velocity Control System uses an INTEGRATING amplifier, as discussed in Chapter 2.

The GAIN of the position control amplifier was expressed in milliamps of output per volt of input. However, in a velocity control amplifier the GAIN is expressed in milliamps per second of output, per volt of input.

In other words, the position control amplifier produced a **constant amount** of current for a given input signal.

However, the velocity control amplifier produces a **changing** amount of current. The given input signal controls **how fast the current output will change**.

Gain in an integrating, or velocity control, amplifier is expressed in ma/sec/volt.

The units of gain therefore state how fast the output will change, per volt of input.

Since a velocity control system is controlling actuator speed, the feedback transducer for a velocity control system converts the SPEED of the actuator into a feedback voltage.

The open loop gain, K_V, is a direct measurement of HOW FAST the system will **change the speed** of the actuator. If $K_V = 0.2 \, \omega_S$, the load will reach the commanded speed within 5 time constants, where 1 time constant is equal to $1/K_V$.

This is why K_V, in a velocity control system, is sometimes referred to as the ACCELERATION CONSTANT. Increasing the gain will increase the acceleration of the actuator.

Open loop gain, in a velocity control system, is further described by referring to it as K_{VV}. The second "V" in the subscript reveals that the gain being discussed is used in a VELOCITY CONTROL SYSTEM – not a position or force control system.

$$K_V \;=\; (G_{AMP})(G_{SV})(G_{CYL})H_{FB}$$

K_V is still expressed in units of sec^{-1}.

$$K_{VV} \;=\; \frac{ma/sec}{volt} \; \frac{in^3/sec}{ma} \; \frac{1}{in^2} \; \frac{volts}{in/sec} \;=\; \frac{1}{sec} \;=\; sec^{-1}$$

When an integrating amplifier is used in the control loop, any error between the command input and the actual speed of the actuator will cause the output of the amplifier to continue increasing, until the error signal is equal to zero.

Therefore, in a velocity control system, **the steady-state errors** due to internal and external disturbances **will be nearly zero!**

The main error induced in a velocity control system comes from the feedback transducer – so the accuracy of the transducer tends to be the limiting factor in the overall accuracy of the system.

A step-input voltage to the velocity control system will control the speed of the actuator.

If control of actuator acceleration and deceleration is desired, a ramp generator can be used to supply a gradually increasing input, as shown in Figure 6.37:

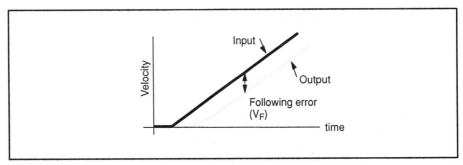

Figure 6.37

When a ramped signal is used in a velocity control system, there will be a FOLLOWING ERROR for the same reason as in a position control system – a mechanical actuator cannot respond instantly to a change.

The size of the following error will depend on the open loop gain of the system, and how fast the system is attempting to change its speed.

$$V_F = \frac{a}{K_{VV}}$$

Equation #50

Where: V_F = Velocity Following Error (inches/sec)
a = Actuator acceleration (inches/sec^2)
K_{VV} = Open Loop Gain (sec^{-1})

Force Control System

Force control systems, as the name implies, control the amount of FORCE generated by the actuator. The term "Force Control" usually implies that a linear actuator, (a cylinder, for example) is being controlled.

When a rotary actuator (such as a motor) is being force controlled, it is more technically correct to call the system a "Torque Control" system, since torque is the actual quantity being controlled.

Force or torque control systems can sense force or torque directly by using a load cell or torque transducer as a feedback device.

However, it is usually more convenient to sense actuator pressure by means of a pressure transducer, since the amount of hydraulic pressure applied to the actuator is a direct indication of the force or torque that will be generated by the actuator.

When sensing actuator pressure, it is important to realize the effect of BACKPRESSURE on the other side of the actuator. Any backpressure on the outlet port will subtract from the force being generated by the pressure applied to the inlet side of the actuator.

Equal area actuators, such as double rod cylinders and motors, can directly use a differential pressure transducer as shown in Figure 6.38:

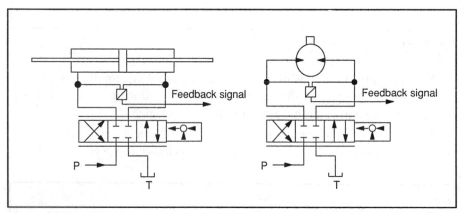

Figure 6.38

For example, assume that the double rod cylinder in Figure 6.38 has a piston area of 10 in^2 on each side.

Assume further that 1000 PSI is applied to the left port of the cylinder, and that the application creates a 100 PSI backpressure on the right port of the cylinder.

The force generated on the left side is:

$$Force \ = \ Pressure \times Area \ = \ 1000 \ \frac{lb}{in^2} \times 10 \ in^2 \ = \ 10,000 \ lb_f$$

However, the backpressure on the right side will generate an opposing force of:

$$Force \ = \ Pressure \times Area \ = \ 100 \ \frac{lb}{in^2} \times 10 \ in^2 \ = \ 1,000 \ lb_f$$

This means that the force actually available from the actuator is:

$$10,000\ lb_f\ -\ 1,000\ lb_f\ =\ 9,000\ lb_f$$

By using a DIFFERENTIAL pressure transducer, which reads the pressure difference across the actuator, a correct force reading can be obtained for feedback purposes:

$$Force\ =\ Differential Pressure \times Area$$
$$Force\ =\ (1,000\ PSI\ -\ 100\ PSI) \times 10\ in^2$$
$$Force\ =\ 900\ PSI \times 10\ in^2\ =\ 9,000\ lb_f$$

Control of force generated by an UNEQUAL area actuator, such as a single rod cylinder, becomes a bit more complicated because of the area difference between the two sides of the actuator (Figure 6.39):

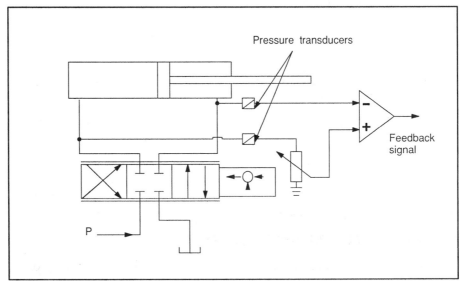

Figure 6.39

One option is to use two identical pressure transducers, and scale the output of one transducer to account for the piston area difference as shown in Figure 6.39.

Differential pressure transducers are also available which can be internally adjusted to accomplish this scaling.

The simplest method of controlling force from an unequal area actuator is to use a single pressure transducer on the driving end of the cylinder, and eliminate the backpressure in the opposite end of the cylinder by connecting the outlet side directly to tank through a solenoid valve.

Of course, the solenoid valve must somehow be opened during the part of the machine cycle when the pressure is being controlled.

The open loop gain, K_V, is a direct measurement of HOW FAST the system will apply the commanded force. If $K_V = 0.2 \, \omega_S$, the actuator will apply the commanded force or torque within 5 time constants, where 1 time constant is equal to $1/K_V$.

Open loop gain, in a force control system, is further described by referring to it as K_{VP}. The "P" in the subscript reveals that the gain being discussed is used in a PRESSURE CONTROL SYSTEM – not a velocity or position control system.

Open loop gain (K_{VP}) in a pressure control system is calculated in exactly the same way as in a position or velocity control system, and is measured in units of 1/sec.

The equation for finding K_{VP} remains:

$$K_{VP} \;=\; (G_{AMP})(G_{SV})(G_{CYL})H_{FB}$$

In a pressure control system, however, the gain of the servo valve (G_{SV}) can be tricky to calculate because its output is now expressed as a RATE OF FORCE INCREASE (lb_f/sec) per ma of input.

This indicates that the valve is being used to regulate PRESSURE rather than FLOW.

The easier method of finding K_{VP} is to first determine the natural frequency of the system, and then calculate K_{VP} using the guidelines previously stated in Case A, B and C. The amplifier gain is then adjusted so that this desired value of K_{VP} is not exceeded.

CALCULATION OF PRESSURE ERROR

Steady-state error in a pressure control system is caused by a variety of factors:

- Leakage in the valve
- Valve uncertainties (hysteresis, threshold, null shift)
- Movement of the actuator
- Accuracy of the feedback transducer
- Leakage in the rest of the system

ERROR DUE TO LEAKAGE IN THE VALVE:

Steady State errors due to leakage in the valve are determined by certain valve characteristics and system parameters. The following equation can be used to estimate this error:

$$\Delta P_L \;=\; 2 \times 10^{-2} \left[\frac{Q_{RP}(C_{TOT})}{(A^2)K_{VP}} \right] \left(\frac{\Delta P_{AB}}{P_S} \right)$$

Equation #51

Where: ΔP_L = Pressure error due to valve leakage (PSI)

Q_{RP} = Rated flow of valve at operating pressure (in^3/sec)

C_{TOT} = Total stiffness of actuator and mounting (lb$_f$/in)

A = Actuator area (in^2)

K_{VP} = Open loop gain (sec^{-1})

ΔP_{AB} = Pressure difference across actuator ports (PSI)

P_S = System supply pressure (PSI)

ERROR DUE TO VALVE UNCERTAINTIES:

Pressure errors due to valve parameters other than leakage, such as hysteresis, threshold, null shift, etc., can be estimated from the following equation:

$$\Delta P_U \;=\; 4 \times 10^{-2} \left[\frac{Q_{RP}(C_{TOT})}{(A^2)K_{VP}} \right]$$

Equation #52

Where: ΔP_U = Pressure error due to valve uncertainties (PSI)

ERROR DUE TO ACTUATOR MOTION:

During the closed loop control of pressure, actuator motion will necessarily contribute to the steady state error of the system.

In order for flow to exist through the valve, the spool must be displaced from the null position.

In order to have the spool displaced from the null position, an error signal **MUST** exist. This error, necessary to produce actuator movement, can be determined from the expression:

$$\Delta P_V = V \left(\frac{(C_{TOT})}{(A)K_{VP}} \right)$$

Equation #53

Where: ΔP_V = Pressure error due to actuator velocity (PSI)
　　　　V = Actuator velocity (in/sec)
　　　　　　(Including compensation for actuator leakage)

Note that the velocity measurement specifies "compensation for actuator leakage." The valve spool must be displaced from null position by an error signal. Displacing the valve spool permits flow to the actuator, which will cause the actuator to move at a certain speed. In reality, however, a certain amount of flow will be lost to leakage in:

The actuator itself, or

In auxiliary devices installed between the servo valve and the actuator (e.g. crossline relief valves, solenoid valves, loose fittings, etc).

If this leakage didn't exist, the actuator could move even faster. The actual velocity of the actuator therefore doesn't absolutely represent the amount of flow through the servo valve.

To account for this real-world "short-changing," either a calculation or a measurement of the total leakage should be made.

This leakage flow should then be applied to the actuator on paper, in order to calculate how much velocity the leakage would have produced in the actuator.

This "lost velocity" due to leakage should then be added to the actual velocity, producing the compensated velocity needed for the velocity error calculation.

ERROR DUE TO THE FEEDBACK TRANSDUCER:

This error is derived from the data sheet or catalog information on the transducer being used.

ΔP_{FB} is typically expressed as a percentage of "span," which is simply a percentage of the pressure range that it was designed to measure.

For example, a pressure transducer which can measure from 0 to 1500 PSI, with an accuracy of 5% of span, yields a maximum error of:

$$1500 \, PSI \times 5\% = 1500 \, PSI \times .05 = \pm 75 \, PSI \, error$$

TOTAL PRESSURE ERROR

The total pressure error can now be found by simply adding up the errors produced by valve leakage, valve uncertainty, actuator motion and feedback transducer limitations:

$$\Delta P_{TOT} \;=\; \Delta P_L + \Delta P_U + \Delta P_V + \Delta P_{FB}$$

Equation #54

Where: ΔP_{TOT} = Total Steady-State Pressure Error

PRESSURE FOLLOWING ERROR

When a changing input signal is applied, pressure systems exhibit a following error, just as position and velocity control systems do.

Figure 6.34 showed the following error in a position control system, and Figure 6.37 illustrated following error in a velocity control system.

To review, following error occurs because of the time lag between the application of the input signal and the actual response of the system to that signal.

Pressure following error can be calculated using:

$$P_f \;=\; \frac{\frac{\Delta P}{\Delta t}}{K_{VP}}$$

Equation #55

Where: P_f = Pressure Following Error (PSI)

$\Delta P/\Delta t$ = Rate of change of pressure (PSI/sec)

K_{VP} = Open Loop Gain (sec^{-1})

An example calculation of pressure error in a force control system is now performed. It is important to list the required application data and do the calculation in organized steps:

1. Find the Hydraulic Stiffness of the System (C_H)
2. Use C_H to find the Load Natural Frequency (ω_L)
3. Use ω_L and ω_V to find System Natural Frequency (ω_S)
4. Use ω_S to find Open Loop Gain (K_{VP})
5. Calculate valve rated flow at system pressure (Q_{RP})

6. Use K_{VP} and Q_{RP} to find individual errors due to
    a. Valve Uncertainties
    b. Actuator Movement
    c. Valve Leakage
    d. Feedback Transducer Errors
7. Add up the individual errors to find P_{TOT}.

Example 6.10:

Calculate the total steady-state pressure error for the system shown in Figure 6.40:

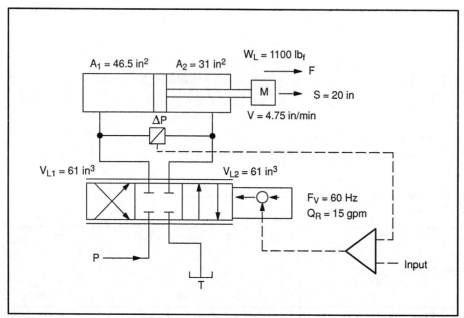

Figure 6.40

The pressure feedback transducer is internally scaled to compensate for the cylinder area ratio, and has a 0-1500 PSI span with 1% accuracy. The valve has a symmetrical spool.

STEP 1: Find the Hydraulic Stiffness of the system.

$$C_H = E \left[\frac{(A_1)^2}{V_{L1} + V_1} + \frac{(A_2)^2}{V_{L2} + V_2} \right]$$

Assume that minimum stiffness occurs at midstroke.

$$V_1 = \frac{S}{2}(A_1) \text{ and } V_2 = \frac{S}{2}(A_2)$$

$$C_H = 2 \times 10^5 \left[\frac{46.5^2}{61 + \frac{20}{2}(46.5)} + \frac{31^2}{61 + \frac{20}{2}(31)} \right] = 1.34 \times 10^6 \; lb_f/inch$$

STEP 2: Find the natural frequency of the load

$$\omega_L = \sqrt{\frac{C_H}{M}}$$

We know the WEIGHT of the load, but need to know its MASS.

Since *Weight* $= \; Mass \times Gravity$, then $M = \dfrac{W}{g}$

$$M = \frac{1100 \; lb}{386 \; in/sec^2} = 2.85 \frac{lb \; sec^2}{inch}$$

$$\omega_L = \sqrt{\frac{1.34 \times 10^6}{2.85}} = 685.7 \; radians/sec$$
$$so \; .3\omega_L = 205.7 \; radians/sec$$
$$and \; 3\omega_L = 2057 \; radians/sec$$

STEP 3: Determine the natural frequency of the system

From Figure 6.40, the valve's natural frequency is

$$f_V = 60Hz$$

Since $\omega = 2\pi f$, this translates to

$$\omega_V = 2 \times \pi \times 60 = 377 \; radians/sec$$

Use Case B to calculate system natural frequency, since:

$$.3\omega_L < \omega_V < 3\omega_L$$
$$205.7 < 377 < 2057$$

Therefore,

$$\omega_S = \frac{\omega_L(\omega_V)}{\omega_L + \omega_V} = \frac{685.7(377)}{685.7 + 377} = 243.3 \; radians/sec$$

STEP 4: Find the open loop gain of the system

$$K_{VP}(max) = .2\omega_S = .2(243.3) = 48.65 \; sec^{-1}$$

STEP 5: Find valve rated flow at operating system pressure

$$Q_{RP} = Q_R \sqrt{\frac{P_S}{1000}}$$

The valve is currently rated at $Q_R = 1$ GPM. We need to convert Q_R to cubic inches per second for this calculation.

$$Q_R = 1 \frac{gallon}{minute} \times 231 \frac{in^3}{gallon} \times \frac{1\ minute}{60\ seconds} = 3.85 \frac{in^3}{sec}$$

Q_R is now expressed in proper units for calculating Q_{RP}:

$$Q_{RP} = 3.85 \sqrt{\frac{1450}{1000}} = 4.64\ in^3/sec$$

STEP 6: Calculate individual pressure error factors

a.) Calculate ΔP_U – error due to valve uncertainties.

$$\Delta P_U = 4 \times 10^{-2} \left(\frac{Q_{RP}(C_{TOT})}{(A^2)K_{VP}} \right)$$

$$\Delta P_U = 4 \times 10^{-2} \left(\frac{4.64 \times 1.34 \times 10^6}{46.5 \times 46.5 \times 48.65} \right) = 2.36\ PSI$$

b.) Calculate ΔP_V – error due to actuator speed

$$\Delta P_V = V \left(\frac{C_{TOT}}{(A)K_{VP}} \right)$$

Velocity V is expressed as 4.75 inches per minute. The units required by the above equation are inches per SECOND.

Therefore,

$$\Delta P_V = \frac{4.75\ in}{min} \times \frac{1\ min}{60\ sec} \times \left[\frac{1.34 \times 10^6}{46.5 \times 48.65} \right] = 46.9\ PSI$$

c.) Calculate ΔP_L – error due to leakage in the valve

$$\Delta P_L = 2 \times 10^{-2} \left(\frac{Q_{RP}(C_{TOT})}{(A^2)K_{VP}} \right) \left(\frac{\Delta P_{AB}}{P_S} \right)$$

All of the above quantities are known except ΔP_{AB}, which must be calculated.

The system is designed to supply a constant 1100 lb. of actual output force at a speed of 4.75 inches per minute (or about .080 inches per second). Going back to Equation #5,

$$F_{TOT} = F_a + F_c + F_E + F_S$$

The external force, F_E, is assumed to be 0.

The acceleration force, F_a, is very small and can be ignored in this application, since the cylinder "creeps" along very slowly.

The 1100 lb_f requirement can be treated as a Frictional force, F_c.

The Cylinder Seal Friction Force is assumed to be 1/10 of F_{TOT}.

Therefore,

$$F_{TOT} = F_a + F_c + F_E + F_S$$
$$F_{TOT} = 0 + 1100 + 0 + .1 F_{TOT}$$
$$.9 F_{TOT} = 1100$$
$$F_{TOT} = 1222 \, lb_f$$

Equations 6 and 7 can now be used to find P_1 and P_2:

R is the cylinder area ratio, equal to A_1/A_2

$$R = \frac{46.5}{31} = 1.5$$

By Equation #6,

$$P_1 = \frac{P_S(A_2) + R^2 [F + P_T(A_2)]}{A_2 (1 + R^3)}$$

$$P_1 = \frac{1450(31) + 1.5^2 [1222 + 0(31)]}{31 (1 + 1.5^3)}$$

$$P_1 = \frac{44,950 + 2,749.5}{135.625} = 351.7 \, PSI$$

By Equation #7,

$$P_2 = P_T + \frac{P_S - P_1}{R^2} = 0 + \frac{1450 - 351.7}{2.25} = 488.1 \, PSI$$

ΔP_{AB} can now be calculated:

$$\Delta P_{AB} = P_2 - P_1 = 488.1 - 351.7 = 136.4 \, PSI$$

$$\Delta P_L = 2 \times 10^{-2} \left(\frac{Q_{RP}(C_{TOT})}{(A^2)K_{VP}} \right) \left(\frac{\Delta P_{AB}}{P_S} \right)$$

$$\Delta P_L = 2 \times 10^{-2} \left(\frac{4.64 \times 1.34 \times 10^6}{46.5 \times 46.5 \times 48.65} \right) \left(\frac{136.4}{1450} \right) = 0.1 \, PSI$$

d.) Find ΔP_{FB} – error induced by the feedback transducer

As stated earlier, the pressure feedback transducer is internally scaled to compensate for the cylinder area ratio, and has a 0-1500 PSI span with 1% accuracy.

$$\Delta P_{FB} = 1500 \, PSI \times 0.01 = 15 \, PSI \; maximum \; error$$

STEP 7: Calculate the total pressure error – ΔP_{FB}

$$\Delta P_{TOT} = \Delta P_L + \Delta P_U + \Delta P_V + \Delta P_{FB}$$

$$\Delta P_{TOT} = .11 + 2.36 + 46.9 + 15 \, PSI = 64.37 \, PSI$$

This system, as defined, will control pressure to ±64.4 PSI as the cylinder extends.

If the load is stopped during extension of the cylinder, the cylinder will continue to apply 1100 lb of force to the load.

However, since V = 0, the error due to cylinder motion will disappear, and

$$\Delta P_{TOT} = .11 + 2.36 + 0 + 15 \, PSI = \pm \, 17.47 \, PSI$$

The feedback transducer now becomes the major source of pressure error.

More Advanced Control Techniques

So far, only simple control amplifiers have been considered.

These include the proportional amplifiers used in position control systems, and the integrating amplifiers described in connection with velocity control systems.

Addition of integrating capability to a proportional amplifier creates what is called a PROPORTIONAL plus INTEGRAL amplifier (P+I type), which can be used to reduce steady-state error in both velocity and force control systems.

In addition to the proportional and integral terms, there is also a term known as the DERIVATIVE term.

This DERIVATIVE term is a gain which is related to the RATE OF CHANGE of the error signal. If it is available in a given control amplifier, it can be used to make the system respond faster without necessarily causing overshoot or oscillation.

Before the effects of these terms can be clearly explained, it is necessary to understand that they are based upon, and affected by, the actual motion of the system.

The actual motion is composed of POSITION, VELOCITY (speed) and ACCELERATION effects. These physical effects are all related to each other.

The starting point, then, in understanding PID (proportional + integral + derivative) systems is to first understand how position, velocity and acceleration are related to each other.

Position, Velocity and Acceleration

Position of the actuator is an easy concept to see and understand. It is simply the location of the load at any given instant in time.

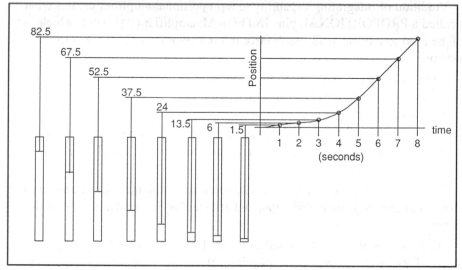

Figure 7.1

The above graph shows the POSITION of a moving cylinder. Specific position measurements are shown at eight different times.

The curve illustrates how the cylinder changes position over an eight second interval.

If enough is known about how the cylinder changes position, the VELOCITY (or speed) of the cylinder can also be plotted.

VELOCITY can be viewed as the cylinder's RATE OF CHANGE OF POSITION. It tells how fast the cylinder is changing position.

Figure 7.2 shows the velocity of this cylinder plotted beneath the position curve.

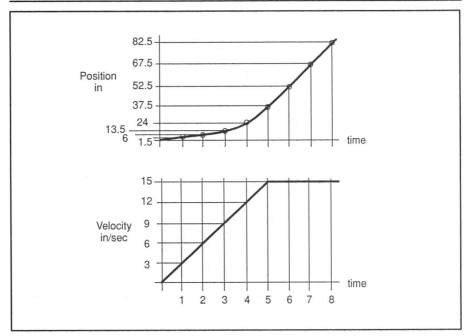

Figure 7.2

As you can see, the cylinder is speeding up during the 0 to 5 second time interval, then holding a constant speed of 15 in/sec during the last 3 seconds.

There are two important sections of these curves which deserve another look – The 0 to 5 second, and the 5 to 8 second intervals.

During the 0 to 5 second time interval, the velocity is constantly increasing. This constant RATE OF CHANGE of velocity is shown as a straight, sloping line on the velocity plot.

During this period of changing velocity, notice that the POSITION plot is **curved**. This is called an EXPONENTIAL curve. In other words, the position is changing more and more rapidly as time goes on. With each passing millisecond, the cylinder is moving faster and faster.

However, after 5 seconds, the cylinder reaches it's maximum speed of 15 in/sec. The Velocity curve flattens out, causing the Position curve to change to a straight, sloping line. This indicates that the position is now changing at a CONSTANT RATE.

Velocity, therefore, is HOW FAST THE POSITION IS CHANGING.

ACCELERATION is the RATE OF CHANGE OF VELOCITY. This concept is shown in Figure 7.3:

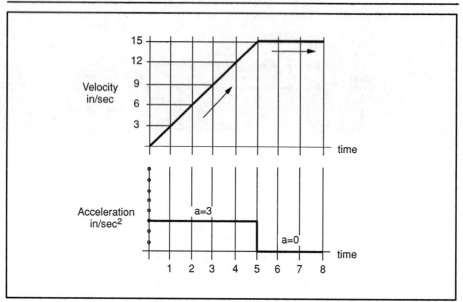

Figure 7.3

Once again, focus on the 0 to 5 second and the 5 to 8 second time intervals.

During the 0 to 5 second time interval, the cylinder is constantly speeding up. The velocity is constantly changing.

The graph shows that after 1 second, the velocity is 3 in/sec. After 2 seconds, the velocity is 6 in/sec. After 3 seconds, it is 9 in/sec, and so on.

In other words, the velocity is changing at a CONSTANT RATE of 3 in/sec, every second – or 3 in/sec^2.

From 5 to 8 seconds, the velocity is a constant 15 in/sec. The cylinder is still moving and still changing position, but it is doing so at a constant speed.

Since the speed isn't CHANGING, its ACCELERATION IS ZERO. It is not accelerating anymore.

To recap, VELOCITY is the RATE OF CHANGE OF POSITION and ACCELERATION is the RATE OF CHANGE OF VELOCITY.

Derivatives and Integrals

DERIVATIVES:

Looking at the previous position, velocity and acceleration plots for the cylinder, you can see that the control system is now concerned with RATES OF CHANGE.

Velocity, or speed, is nothing more than the RATE OF CHANGE OF POSITION.

Acceleration is simply the RATE OF CHANGE OF VELOCITY.

There are specific terms used when talking about these rates of change.

Origin of the words "Derivative" and "Integral":

CALCULUS is a branch of mathematics that was created and developed over the centuries to accurately deal with **rates of change**. While we will not get deeply into calculus within the scope of this text, we will need to refer to some of its basic concepts.

The term "Derivative," for example, is a calculus term that means "The Rate of Change of."

For example, Velocity is the Derivative of (or Rate of Change of) Position, Acceleration is the Derivative of (or Rate of Change of) Velocity, and, Acceleration is therefore a DOUBLE DERIVATIVE of (or the Derivative of the Derivative of) Position.

The ability to calculate derivatives requires several semesters of college calculus – but fortunately, we can understand the effect of DERIVATIVE ACTION in a control system by looking at a simpler underlying concept – the TANGENT LINE.

THE TANGENT LINE:

Figure 7.4 shows the first five seconds of the cylinder's motion.

During this five seconds, the cylinder is accelerating at a constant rate of 3 in/sec^2. During this time, both the position and velocity are changing.

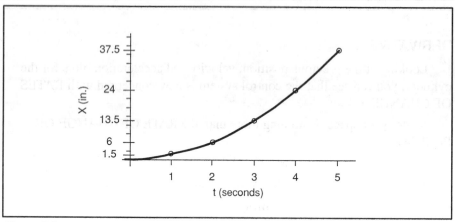

Figure 7.4

The graph shows the exact position of the cylinder at 5 different times. For example, at 3 seconds the cylinder has moved to the 13.5 inch position.

Exactly how fast is the cylinder moving at t = 3 seconds?

If you know the mathematical equation that describes the above curve, and know how to find its derivative, you could calculate it – like this:

Equation of Curve is $x = \dfrac{3t^2}{2}$, so $v = \dfrac{dx}{dt} = \dfrac{3}{2}(2t^1) = 3t$

Velocity at t = 3 is: $v = 3t = 3(3) = 9 \; in/sec$

Using the calculated derivative (V = 3t), you can find the cylinder's velocity at any position as it moves.

HOWEVER, THERE IS AN EASIER WAY!

The velocity of the cylinder can be found at any point by drawing a tangent line to the curve at that point. This is illustrated by Figure 7.5:

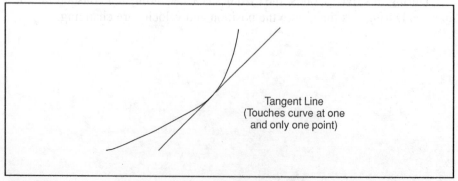

Tangent Line
(Touches curve at one
and only one point)

Figure 7.5

A "tangent line" is a straight line that touches the curve at only one point.

This line, once placed on the curve, has a "SLOPE." The slope of the line is the velocity of the cylinder at that position.

For example, Figure 7.6 shows the "slope" of the tangent line drawn to the point t = 1 second.

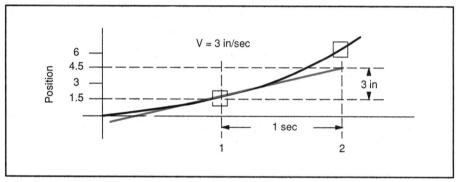

Figure 7.6

The "slope" of the line is found by dividing its RISE by its RUN.

Since the line is straight, it doesn't matter how long a section of the line you use, so pick any two section ends that are convenient.

In Figure 7.6 we have chosen two points on the line, that cross at t = 1 and t = 2. The positions at these two points are x = 1.5" and x = 4.5".

$$Slope\ of\ the\ line \;=\; \frac{RISE}{RUN} \;=\; \frac{(4.5'' - 1.5'')}{(2\ sec - 1\ sec)}$$

Over one second, this line "rises" 3 inches, indicating a speed of 3 inches per second. This gives the same result as the calculated derivative:

$$V \;=\; 3t \;=\; 3(1) \;=\; 3\ in/sec$$

Looking at t = 2 seconds (Figure 7.7), we see that the velocity has increased.

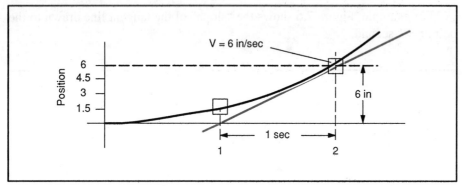

Figure 7.7

$$Slope = \frac{RISE}{RUN} = \frac{(6'' - 0'')}{(2\ sec - 1\ sec)} = 6\ in/sec$$

At t = 3 seconds, Figure 7.8 shows an even greater slope, indicating that velocity continues to increase.

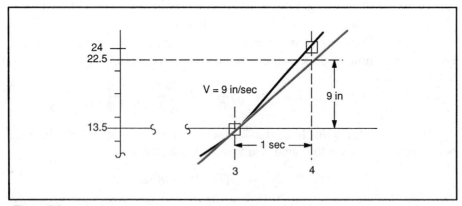

Figure 7.8

$$Slope = \frac{RISE}{RUN} = \frac{(22.5'' - 13.5'')}{(4\ sec - 3\ sec)} = 9\ in/sec$$

Comparing the slopes of the tangent lines in Figures 7.6, 7.7, and 7.8, the increasing velocity becomes visible as the position of the cylinder changes.

The DERIVATIVE of the position (the velocity) starts out small when the cylinder first begins to move, and gets increasingly larger as the cylinder picks up speed.

Eventually, at t = 5 seconds, the velocity hits maximum and levels off at 15 in/sec. The position "curve" becomes a straight line. At this point, the slope of the tangent line becomes the same as the slope of the position "curve," and it does not increase any further.

INTEGRATION:

Most math processes are "reversible." What you do mathematically can be undone mathematically.

This is true in Calculus. The opposite of the DERIVATIVE is the INTEGRAL, and the opposite of "taking a derivative" is called "INTEGRATION" or "INTEGRATING."

Where a derivative is useful in finding out how fast something is changing, the integral is useful in taking a changing quantity and totalling up its end effects.

For example, by finding the derivative of the cylinder's position we can calculate the rate of change of position – which we call speed or velocity. In other words, we can use position to find velocity.

But by INTEGRATING the velocity over a set time period, we can find out how far the cylinder moved! In other words, we can use velocity to find position – the reverse process.

VELOCITY is an integral of ACCELERATION, and POSITION is an integral of VELOCITY

The simplest way to understand the process of integration is to view an example (Figure 7.9):

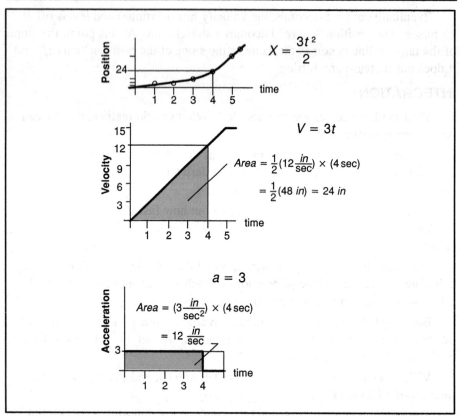

Figure 7.9

The cylinder undergoes a constant acceleration of 3 in/sec^2. What is the position of the cylinder after 4 seconds?

The process of Integrating a function basically involves summing up all of the infinitely small changes caused by the function to find their total effect.

In our example, this can also be done by finding the "area under the curve."

In Figure 7.9, the acceleration from 0 to 4 seconds can be integrated by finding the area under the Acceleration plot between 0 and 4 seconds.

The area covered is:

$$3 \frac{in}{sec^2} \times 4 \; seconds \;\; = \;\; 12 \; in/sec$$

The velocity at t = 4 seconds is 12 in/sec. This can be verified by looking at the Velocity plot in Figure 7.9. The velocity at t = 4 is indeed 12 in/sec.

By integration, acceleration has been used to find velocity.

If the same process is repeated using the Velocity plot, the position of the cylinder at t = 4 seconds can be found.

At t = 4 seconds, the velocity is 12 in/sec. Notice, however, that the "area under the curve" is triangular – it is a square, cut in half.

The area under this curve can be found by finding the area of the square (12 in/sec x 4 sec) and taking HALF of it.

The area covered is:

$$\frac{1}{2}(12\ in/sec \times 4\ sec)\ =\ \frac{1}{2}(48)\ =\ 24\ in$$

By integration, the cylinder extends 24 inches after 4 seconds. Velocity has been used to find position.

These DERIVATIVE and INTEGRAL concepts are used by advanced motion controllers, to minimize response time and increase accuracy.

Figure 7.10 illustrates the effect of increasing the gain in a conventional position control system which is responding to a step input, using only proportional gain.

Proportional Plus Derivative Control (P+D)

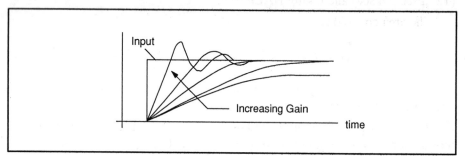

Figure 7.10

Increasing the gain causes the speed of response to increase. But beyond a certain point, OVERSHOOT begins to occur. Increasing the gain even further will only make the overshoot worse, and will also increase the settling time.

It would be VERY desirable to increase the proportional gain and improve the system's speed of response, but to somehow reduce or eliminate the overshoot at the same time.

This is the purpose of DERIVATIVE GAIN, which is achieved by adding a DERIVATIVE TERM to the amplifier (Figure 7.11):

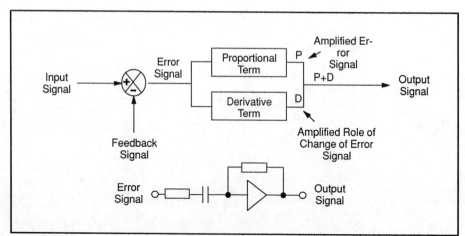

Figure 7.11

Derivative capability, in its simplest form, can be added to a proportional amplifier by installing a properly sized capacitor to the input as shown.

Functionally, the error signal undergoes the proportional amplification of the amplifier, and a derivative amplification which is controlled by the RATE OF CHANGE OF THE ERROR SIGNAL.

The derivative gain tends to SUBTRACT from the proportional gain, and act as a "brake" on the error signal.

When the error signal is changing slowly (for example, at the beginning of the motion), the derivative term is small and has little effect. (Figure 7.12):

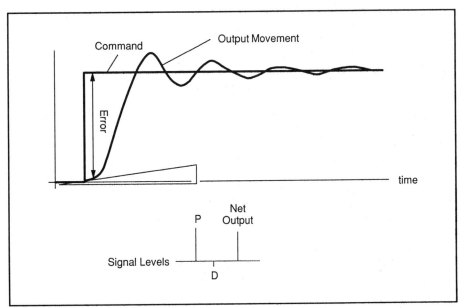

Figure 7.12

The proportional signal is large because the error is large. However, the derivative term is small due to the slow rate of change of the error signal.

The net output (P+D) is therefore large in the "forward" direction – opening the control valve wide.

Things gradually change, however, as the load progresses toward commanded position.

As the load gains velocity, the derivative term gets larger.

At the same time the proportional signal begins to decrease, because the error signal is constantly getting smaller as the load approaches commanded position (Figure 7.13):

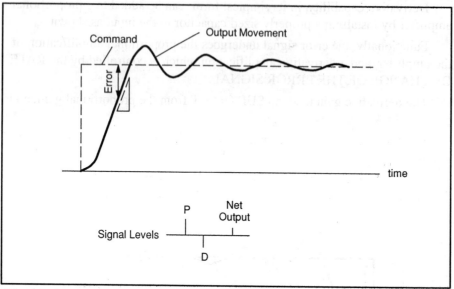

Figure 7.13

This causes the net output of the amplifier to be more rapidly reduced as the commanded position is approached, resulting in more rapid servo valve spool movement back to the null position.

The derivative term therefore causes deceleration of the actuator to begin at an earlier time than would occur with proportional gain alone. This permits a relatively high proportional gain to be used, but avoids the problem of excessive overshoot (Figure 7.14):

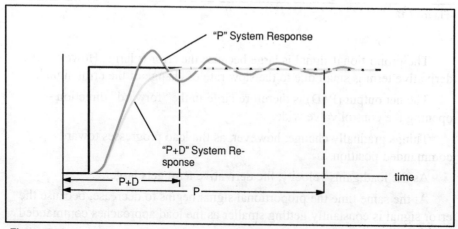

Figure 7.14

The end benefit of derivative gain, when properly adjusted, is to decrease the response time of the system as shown in Figure 7.14.

The ability to increase the proportional gain without causing excessive overshoot has some distinct advantages, one of which is reduction of steady-state error. Increasing K_V, as you will recall, serves to decrease this error.

The derivative term has no effect on steady-state error. If the rate of change of error is zero, then the derivative term is also equal to zero.

Derivative gain, since it acts as a "brake," tends to make a control system more stable. It can be a very useful term to have in an electrohydraulic system, since hydraulic systems tend to have relatively high damping built into them.

It can be a detriment in fast acting systems with low damping, however. If the load approaches commanded signal rapidly enough, resulting in a very rapid rate of change in the error signal, the derivative term can become larger than the proportional term, opening the valve in the other direction.

The signal from the feedback transducer must also be properly shielded to prevent electrical noise on the feedback signal, since the derivative term would respond to the rapidly changing noise.

Integral Control is used to correct for steady-state error, which was previously discussed in Chapter 6.

Steady-state error, as you will recall, is caused by valve uncertainties and external forces acting on the actuator.

Increasing the proportional gain will reduce, but will not eliminate, these errors.

Furthermore, increasing the proportional gain too much may increase overshoot, increase settling time and reduce system stability.

Fortunately, steady-state error can be eliminated through use of an INTEGRAL term in the amplifier.

The integral term comes from an INTEGRATING amplifier. This amplifier is the same type used in a velocity control system or in a ramp generator. A constant voltage on the input produces a gradually increasing voltage on the output (Figure 7.15):

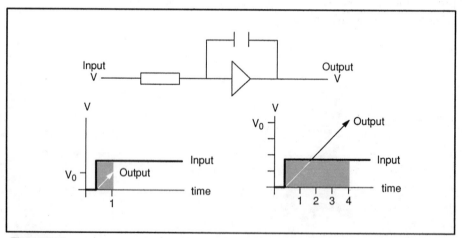

Figure 7.15

This amplifier INTEGRATES its input over time, summing up the area under the input "curve." Figure 7.15 shows the integration process at t = 1 second and t = 4 seconds.

The amplifier "adds up" the area of the curve continuously over time. At t = 1 second, the area under the curve is shown on the left. This results in the output shown.

However, by t = 4 seconds, the area under the curve is 4 times greater, resulting in 4 times more output as shown on the right.

The output of an integrating amplifier is therefore proportional to

Input × Time

If the input happens to be the error signal in a closed loop system then a steady-state error will act in the same way as a fixed input voltage, producing a gradually increasing output as long as the error exists.

By adding an integral term to a proportional amplifier, any steady-state error will now create a gradually increasing drive signal to the control valve. The valve spool will then move sufficiently to drive the actuator and correct the error.

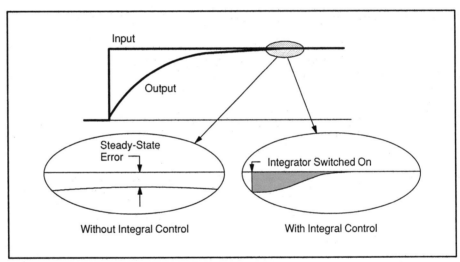

Figure 7.16

Figure 7.16 shows the response of a typical position control system, both with and without integral control.

Without an integral term, the steady-state error produces the condition shown on the left, where the actuator fails to precisely achieve the commanded position.

With an integral term, the steady-state error is "summed up" and output to the valve. The summation stops when the error is zero, causing the integrator output to level off and hold.

The integral term is useful in eliminating **STEADY-STATE ERROR** only.

Steady-state error is the constant error that occurs after the actuator has responded fully to the input, and has "settled."

The integral term becomes a serious disadvantage if it is applied to the DYNAMIC error signal.

The **DYNAMIC** error is the difference between command and feedback **during the motion to commanded position**.

Therefore, use of the integral term for the purpose of eliminating steady-state error requires that the integral term be "switched into the system" ONLY AT THE PROPER TIME.

If the integral term is not controlled in this manner it will happily sum dynamic as well as steady-state error, resulting in serious overshoot and frequently causing loss of stability (Figure 7.17):

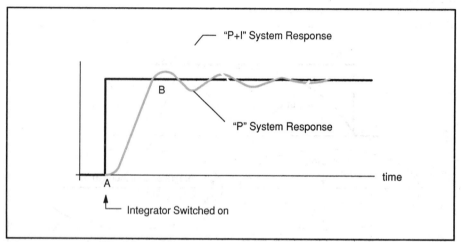

Figure 7.17

This problem can be overcome by either reducing the gain of the integral term, or by switching the integrator off during dynamic conditions (travel to a new position, velocity or force setting.)

The first method is used on simpler P+I controllers that do not have "switchoff" capability. By turning down the integral gain, it may be possible to reduce the overshoot and oscillations to an acceptable level.

This increases both the settling time and the amount of time it takes to integrate the steady-state error.

This may not be a problem in applications where the actuator remains in position for a long period of time (up to several minutes). In these cases, the extra time required to build up the steady-state error correction would be insignificant.

The other alternative is to disable the integrator during actuator motion, and only allow it to operate after the steady-state has been achieved.

A P+I amplifier that accomplishes this is shown in Figure 7.18:

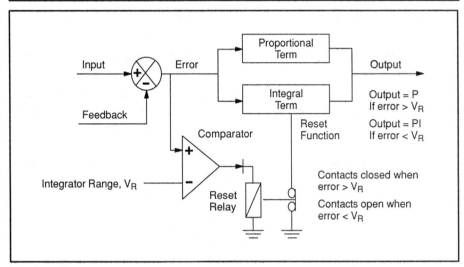

Figure 7.18

The error signal goes to the Proportional and Integral terms, and is also fed to a comparator.

The comparator outputs a positive voltage if the voltage on its positive input is higher than the voltage on its negative input, and it outputs a negative voltage if the positive input is lower than the negative input.

The positive input is connected to the error signal, and the negative input is connected to a reference voltage. The reference voltage is adjusted to a value slightly higher than the error voltage will exhibit at the steady-state.

When the error voltage exceeds the reference, indicating that the system is operating dynamically, the output of the comparator will be a positive voltage.

This positive voltage goes to the RESET relay, via a diode. Since the diode is forward-biased by the positive voltage, electrical current is conducted through the relay coil, energizing the relay and closing the relay contacts.

The closing of the relay contacts DISABLES the integrator, holding it reset so that it cannot operate.

As the system approaches the commanded position (the steady-state), the error signal becomes lower than the reference voltage, causing the comparator to output a negative voltage. The negative voltage reverse-biased the diode, cutting off current to the RESET relay and thus enabling the Integrator to operate.

A number of variations are available in P+I controllers. Some have no reset capability. Others have a RESET input, to which you must apply a voltage or a ground at the proper time – the comparator and relay are not included. More modern controllers use a transistor switch instead of a relay.

The most sophisticated have built-in microprocessors, which continuously analyze the motion profile and optimize the P and I gains.

If a ramp input is applied to a system, FOLLOWING ERROR results, as discussed in Chapter 6.

Increasing the proportional gain will reduce, but not eliminate following error. Adding an Integral term to the control amplifier will eliminate following error, in the same way as with steady-state error (Figure 7.19):

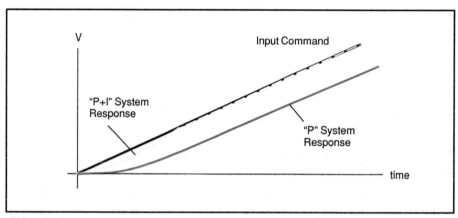

Figure 7.19

Proportional Plus Integral Plus Derivative Control (PID)

The PID controller provides both integral and derivative terms within the control amplifier.

The chief difficulty in using a PID controller is determining the best settings of the Proportional, Integral and Derivative gains.

Merely, "fiddling with them until it works" is a very time consuming method of achieving marginal and unreliable control. It is therefore important to understand the purpose and effects of each of these three terms.

If you are not a Control Systems design expert, it is generally best to approach a PID loop systematically, adjusting one parameter at a time and observing the effects.

Almost everyone, even the expert, reaches a point where the fine adjustment of the loop is done by trial and error.

This section will therefore provide a basic procedure for PID loop tuning, which approaches the adjustments in a step-by-step manner.

LOOP TUNING PROCEDURE:

1. BEFORE ENERGIZING THE CONTROL SYSTEM, review your application carefully. What would happen if your system overshoots? Will damage result?

   In an application where the load moves freely and cannot impact against anything or be crushed by the actuator, overshoot can be tolerated during the adjustment of the loop.

   However in a clamping application, for example, overshoot could damage the actuator or the load being clamped.

   STOP AND ASK YOURSELF – "Will my adjustments cause damage if not done properly?" If so, you may have to modify the equipment for safety reasons, so that it can tolerate over-positioning, overspeed, or higher-than-planned forces during adjustment.

   Common sense will dictate how to do this in most situations.

2. Turn down the P, I and D gains to minimum.

3. Connect a step-input source to the command input of the control amplifier.

4. Verify proper connection of the feedback device to the feedback input of the control amplifier.

   Is the feedback signal properly shielded? If not, the amplifier will amplify the noise and apply it to your control valve.

   Is the feedback signal of the proper polarity? If not, your load will promptly "run-away" when the command signal is applied. This is because the feedback is being added to, not subtracted from, the command signal at the summing junction, producing a constantly increasing error signal.

5. Connect a monitoring device, such as an oscilloscope or an XY
    recorder to the COMMAND INPUT and the FEEDBACK signals.

    Do not attempt to visually "eyeball" the results of the adjustment. Visual
    observation of the actuator is not accurate enough. If you do not know how
    to operate the monitoring device, seek the assistance of someone who
    DOES know how to use it.

6. Energize the system and mechanically null the servo valve.

7. Apply the step input repeatedly, and slowly increase the
    PROPORTIONAL gain of the amplifier a little at a time while
    observing the system response (via the feedback signal).

    A position control system, for example, will respond as shown in
    Figure 7.20:

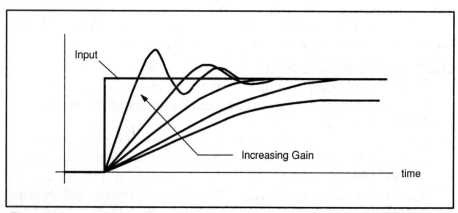

Figure 7.20

As Proportional gain is increased, the step input will give increasingly
faster response. Eventually the gain will reach a value where the input step
will cause the actuator to just barely reach the desired position.

Continue increasing the Proportional gain until the system achieves the
position that the step input should produce, as shown in Figure 7.21:

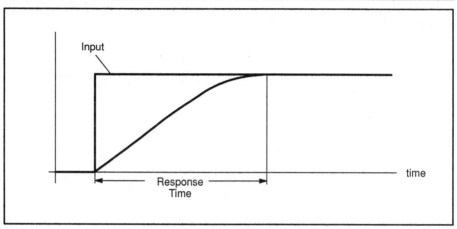

Figure 7.21

8. When the Proportional gain is set high enough so that the step input causes the desired position to be achieved, measure the response time using the time base of monitoring instrument.

9. Now, continue to raise the Proportional gain until a moderate overshoot occurs, as shown in Figure 7.22:

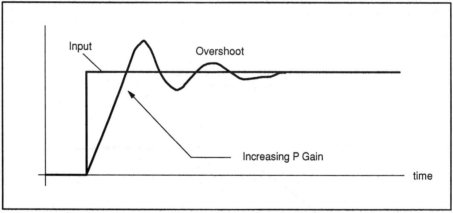

Figure 7.22

10. Begin increasing the DERIVATIVE gain now, a little at a time, while observing the response carefully. The overshoot should begin to reduce as Derivative gain is increased.

   Continue increasing the Derivative gain, and applying the step input, until the overshoot is eliminated (Figure 7.23):

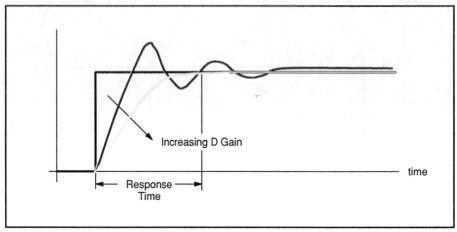

Increasing D Gain

Response
Time

time

Figure 7.23

If you find that you have turned the Derivative gain to maximum, and overshoot is still occurring, then you have set the proportional gain too high. If this occurs, turn the Derivative gain BACK TO MINIMUM, then decrease the Proportional gain a bit and try Step 10 again.

11. When overshoot has been eliminated, measure the response time once again. You should see a marked improvement, as shown in Figure 7.24:

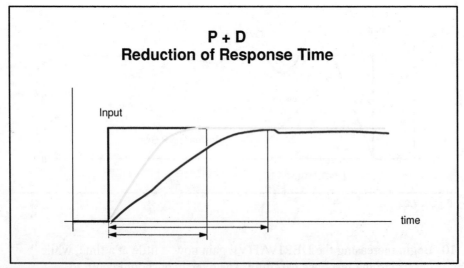

P + D
Reduction of Response Time

Input

time

Figure 7.24

12. When the overshoot is corrected, try to increase the Proportional gain some more, until a slight overshoot occurs. Then see if additional

Derivative gain will correct it. You should soon reach a point where response time cannot be improved any further.

13. Cycle the system several times, using the step input, to see whether the adjustments are stable. If occasional overshoot is noted, decrease the Proportional gain SLIGHTLY, until this stops occurring.

14. When the Derivative gain is set satisfactorily, you can then attempt to set the INTEGRAL gain to eliminate any steady-state error.

    Remember that you want the Integral gain to be disabled (RESET) during the dynamic part of the motion cycle.

    If Integral gain is improperly timed, the overshoots will return. This is usually caused by interaction between the Derivative and Integral terms, as they "fight" each other for control.

    If Integral gain is set too high, you may even lose control of the system as it oscillates out of control. If this occurs, shut down the amplifier, turn the Integral gain back to minimum, and start again.

    When in doubt, set the Integral gain on the LOW side, since it tends to LESSEN THE STABILITY of a closed loop system.

15. Watch the system carefully through the first few days of operation, to ensure that you have not set the gains too high. Some performance "drift" can be expected as the hydraulic fluid changes temperature or as new mechanical components "wear-in," producing the need for some "fine tuning" later on.

Other Control Techniques

There are additional gain terms that you may come across in more sophisticated control devices, or in control devices used on very fast or very slow acting systems.

DUAL GAIN AMPLIFIERS:

Figure 7.25 shows a simple representation of this type of amplifier.

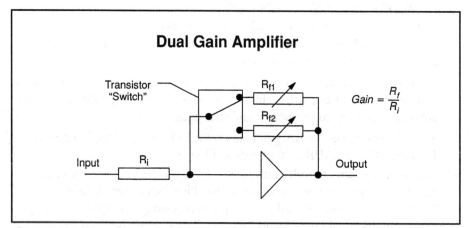

Dual Gain Amplifier

$$Gain = \frac{R_f}{R_i}$$

Figure 7.25

The gain is controlled by the resistance in the feedback loop, divided by the resistance in the input line (Gain = R_f/R_i).

This amplifier, however, has the capability of changing the value of R_f by means of a solid state switch. It can therefore change gain settings automatically.

NON-LINEAR AMPLIFIERS:

Op amps are not limited to just resistors or capacitors in their input or feedback lines.

Up to now, we have viewed the gain of an op amp only as R_f/R_i; but in reality, the gain of an op amp is equal to:

$$Gain \;=\; A \;=\; \frac{Z_f}{Z_i}$$

Z_f and Z_i are frequently just resistors. But they could as easily be simple RL circuits, RC circuits, RLC circuits or complex special purpose circuits that perform some mathematical operation on the gain of the amplifier (Figure 7.26):

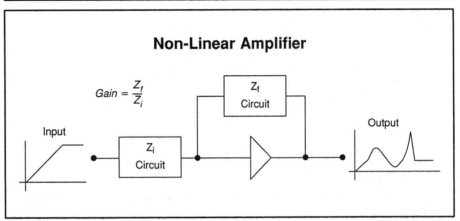

Figure 7.26

This results in a specific GAIN PROFILE where output is not necessarily related directly to the input.

This type of amplifier is generally found in specially engineered systems, and is not typically found in common electrohydraulic applications. They are often not adjustable. Any adjustments that do exist should be left to a qualified control systems specialist who understands the mathematical function being performed.

DOUBLE DERIVATIVE (Kdd) TERMS:

This term is frequently used in low damping, high inertia systems, and is more frequently found in ELECTRIC motion control systems.

The DERIVATIVE TERM, discussed earlier, is developed from the RATE OF CHANGE (Velocity) of the controlled variable. The DOUBLE DERIVATIVE TERM is based on the derivative of the derivative term.

It is therefore affected greatly by the ACCELERATION of the error signal, and its major benefit is to increase the stability of the control system by acting to damp out overshoots rapidly.

The effects of the Double Derivative term are illustrated in Figures 7.27 and 7.28.

Figure 7.27 shows a low inertia system that is overshooting badly.

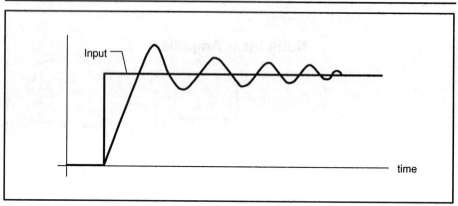

Figure 7.27

Figure 7.28 shows the same system with a Double Derivative term applied.

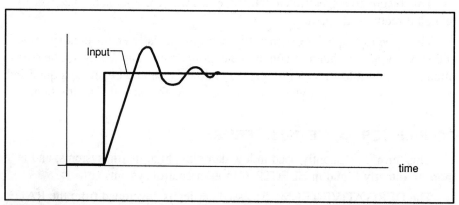

Figure 7.28

The Double Derivative term causes the overshoots to damp out far more rapidly, and also improves the rise time slightly.

When the Double Derivative term is used, it should be adjusted BEFORE adjusting the derivative term (in most controllers). The controller user's manual should be consulted to verify the proper sequence of adjustments.

FEED FORWARD (Kff) GAIN:

This term is generally found in very slowly moving systems or slowly changing processes.

Feed<u>back</u>, by definition, cannot detect and correct an error until the error exists.

Feed <u>Forward</u> gain actually attempts to anticipate the error before it causes an error.

A simple feed forward example is shown in Figure 7.29:

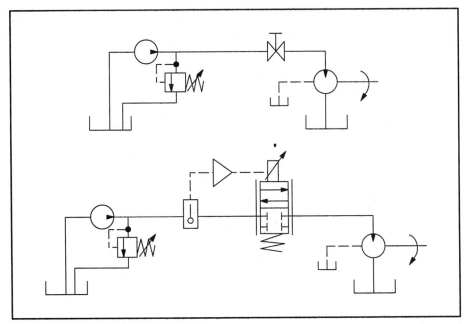

Figure 7.29

At the top of Figure 7.29 is a very simple velocity control for a hydraulic motor which is under a constant, unchanging load.

The pump supplies flow to a manual valve, which is set to control the speed of a hydraulic motor.

This simple control will operate quite well, so long as nothing changes in the system.

However, as this system operates, it is found that the fluid heats up and becomes less viscous, causing the motor speed to change.

The bottom half of Figure 7.29 shows a feed forward fix for this condition. A temperature instrument is placed near the pump outlet.

The temperature instrument outputs a voltage which is amplified and sent to a proportional throttle valve. The system is then calibrated so that the throttle will anticipate any change in viscosity due to the oil temperature, and throttle the motor accordingly.

Feed forward concepts are deceptively simple in appearance, but are usually much more difficult to use in practice.

The simple feed forward on this proportional throttle will react to a temperature change – and nothing else. If motor speed changes due to a change in load, increased internal leakage in the motor, or addition of a different viscosity fluid to the reservoir, the feed forward system cannot react and correct for it.

Therefore, feed forward control is usually used along with feedback control to produce a more reliable system. The feedback system corrects what the feed forward system can't handle.

When a feed forward controller is encountered in a system, the most important consideration is to understand exactly what parameter is being "fed forward" (i.e. temperature, flow, pressure, viscosity, etc.), and to what limits that parameter can affect the system.

Worked Examples

This chapter follows through by providing examples of real-world applications and actual use of the information provided in this text.

Three separate applications are described:

Example 8.1: Determining whether the positional accuracy specification can be met for a hydraulically positioned saw blade.

Example 8.2: Sizing a servo valve and predicting the necessary amplifier gain for a hydraulic velocity control system used to test a jet fighter's landing gear.

Example 8.3: Determining whether an inexpensive proportional type amplifier, or a more costly P+I amplifier will be required in order to meet the specified pressure control requirements on a hydraulic riveting machine.

Example 8.1:

A circular saw blade is used to cut boards of ten different widths, as shown in Figure 8.1:

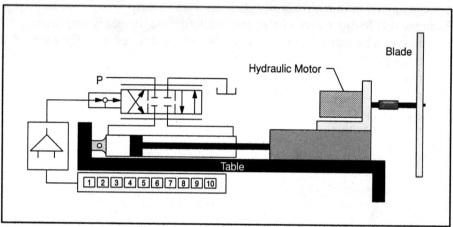

Figure 8.1

The blade is driven by a hydraulic motor.

The motor is bracketed to a sliding carriage, which is moved across the table and positioned by a hydraulic cylinder.

The cylinder position is to be controlled by a Servo valve. The electronic controls for the valve allow the operator to adjust the saw blade to one of ten predetermined positions by selecting the appropriate pushbutton.

The specification for this system requires that the saw blade must be positioned to an accuracy of 0.040 inch (1 millimeter).

DETERMINE:

1. What is the smallest valve that can meet specifications?
2. Can the accuracy specification be met?

ENGINEERING SPECIFICATIONS:

Cylinder Bore Diameter . . . . 1.500 in

Cylinder Rod Diameter . . . . 1.000 in

Cylinder Stroke Length . . . . 25.60 in

Cylinder Seal Friction Force . . . . 16.70 lb_f

Load Weight (Blade, Motor & Carriage) . . . . 660 lb

Pipe Volume (full bore side) . . . . 1.00 in^3

Pipe Volume (rod side) . . . . 13.40 in^3

Friction Coefficient μ (Carriage to Table) 0.15

Maximum Velocity . . . . 11.80 in/sec

Maximum Acceleration . . . . 3.3 ft/sec^2

Hydraulic Supply Pressure . . . . 1740 PSI

Return (Tank) Line Pressure . . . . 44 PSI

Bulk Modulus of Fluid . . . . 2×10^5 lb_f/in^2

Reactive (External) Forces on Carriage . . . . negligible

Actuator Mounting Backlash . . . . negligible

Feedback Transducer Error . . . . negligible

Required Positional Accuracy . . . . .040 in

1. DETERMINE TOTAL FORCE (F)

Find the Total Force required to move the load in the specified manner:

$$\frac{Total}{Force} = \frac{Acceleration}{force} + \frac{Seal}{Friction}_{force} + \frac{Load}{Friction}_{force} + \frac{External}{force}$$

$$F = F_a + F_s + F_c + F_E$$

$$F_a = \frac{W}{386} \times a$$

$$F_a = \frac{W}{386} \times \left(3.3 \frac{ft}{sec^2} \times 12 \frac{in}{ft}\right)$$

$$F_a = \frac{660}{386} \times 39.6 \frac{in}{sec^2}$$

$$F_a = 1.71 \, lb_m \times 39.6 \frac{in}{sec^2}$$

$$F_a = 67.7 \, lb_f$$

$$F_s = 16.7 \, lb_f \text{ (from specifications)}$$

$$F_c = W \times \mu$$
$$F_c = 660 \, lb_f \times 0.15$$
$$F_c = 99.0 \, lb_f$$

$$F_E = 0 \text{ (from specifications)}$$

TOTAL FORCE IS:

$$F = F_a + F_s + F_c + F_E$$
$$F = 67.7 + 16.7 + 99.0 + 0 \, lb_f$$
$$F = 183.4 \, lb_f$$

2. DETERMINE CYLINDER EXTENSION PRESSURES

Find P_1 (full bore) and P_2 (rod end) pressures that will occur during cylinder extension:

Full bore area, A_1

$$A_1 = \frac{\pi \times D^2}{4}$$

$$A_1 = \frac{\pi \times (1.5 \ in)^2}{4}$$

$$A_1 = 1.8 \ in^2$$

Area of the Rod itself, A_r

$$A_r = \frac{\pi \times D^2}{4}$$

$$A_r = \frac{\pi \times (1.0 \ in)^2}{4}$$

$$A_r = .8 \ in^2$$

Area of Rod End, A_2

$$A_2 = A_1 - A_r$$

$$A_2 = 1.8 \ in^2 - .8 \ in^2$$

$$A_2 = 1.0 \ in^2$$

Area Ratio, R

$$R = \frac{A_1}{A_2}$$

$$R = \frac{1.8 \ in^2}{1.0 \ in^2}$$

$$R = 1.8$$

Full Bore Pressure, P_1

$$P_1 = \frac{P_s A_2 + R^2 (F + P_T A_2)}{A_2 (1 + R^3)}$$

$$P_1 = \frac{(1740 \times 1.0) + 1.8^2 (183.4 + [44 \times 1])}{1.0 (1 + 1.8^3)}$$

$$P_1 = 363 \, PSI$$

Rod End Pressure, P_2

$$P_2 = P_T + \frac{P_s - P_1}{R^2}$$

$$P_2 = 44 + \frac{1740 - 363}{1.8^2}$$

$$P_2 = 469 \, PSI$$

3. **DETERMINE RATED FLOW FOR EXTENDING STROKE**
Find the rated flow of the valve (Q_R), that would be required under the conditions existing during cylinder extension:

Actual Flow Required, Q_A

$$Q_A = V_{max} \times A_1$$

$$Q_A = 11.80 \, \frac{in}{sec} \times 1.8 \, in^2$$

$$Q_A = 21.2 \frac{in^3}{sec}$$

Valve's Rated Flow, Q_R

$$Q_R = Q_A \sqrt{\frac{500}{P_s - P_1}}$$

$$Q_R = 21.2 \sqrt{\frac{500}{1740 - 363}}$$

$$Q_R = 12.8 \frac{in^3}{sec}$$

Convert Q_R to GPM

$$Q_R = 12.8 \frac{in^3}{sec} \times \frac{1 \, gallon}{231 \, in^3} \times \frac{60 \, sec}{1 \, minute}$$

$$Q_R = \frac{12.8 \times 60}{231} \frac{gallon}{minute}$$

$$Q_R = 3.3 \, GPM \text{ (minimum for extension)}$$

4. DETERMINE CYLINDER RETRACTION PRESSURES
    Find P_1 (full bore) and P_2 (rod end) pressures that will occur during cylinder retraction:

From step #2:

$$A_1 = 1.8\ in^2$$
$$A_2 = 1.0\ in^2$$
$$R = 1.8$$

Rod End Pressure, P_2

$$P_2 = \frac{P_s A_2 R^3 + F + P_T A_2 R}{A_2(1 + R^3)}$$

$$P_2 = \frac{(1740 \times 1 \times 1.8^3) + 183.4 + (44 \times 1 \times 1.8)}{1^2 \times (1 + 1.8^3)}$$

$$P_2 = 1524\ PSI$$

Full Bore Pressure, P_1

$$P_1 = P_T + R^2 (P_s - P_2)$$
$$P_1 = 44 + 1.8^2 (1740 - 1524)$$
$$P_1 = 744\ PSI$$

---

5. DETERMINE RATED FLOW FOR RETRACTING STROKE
    Find the rated flow of the valve (Q_R), that would be required under the
    conditions existing during cylinder retraction:

Actual Flow Required, Q_A

$$Q_A = V_{max} \times A_2$$
$$Q_A = 11.80 \frac{in}{sec} \times 1.0 \ in^2$$
$$Q_A = 11.8 \frac{in^3}{sec}$$

Valve's Rated Flow, Q_R

$$Q_R = Q_A \sqrt{\frac{500}{P_s - P_2}}$$
$$Q_R = 11.8 \sqrt{\frac{500}{1740 - 1524}}$$
$$Q_R = 18.0 \frac{in^3}{sec}$$

Convert Q_R to GPM

$$Q_R = 18.0 \frac{in^3}{sec} \times \frac{1 \ gallon}{231 \ in^3} \times \frac{60 \ sec}{1 \ minute}$$
$$Q_R = \frac{18.0 \times 60}{231} \frac{gallon}{minute}$$
$$Q_R = 4.7 \ GPM, \text{(minimum for retraction)}$$

Choice of Valve Size:

Q_R required for extension is: 3.3 GPM

Q_R required for retraction is: 4.7 GPM

Minimum rated flow of valve to be used for this application is
therefore: 4.7 GPM

The smallest valve in Appendix E (page 301, Model Code info)
meeting spec is: SM4 – 10(5)19 rated for 5 GPM (19 l/min)

6. DETERMINE CYLINDER STROKE FOR MINIMUM STIFFNESS

Find the point in the cylinder's stroke at which minimum hydraulic stiffness is experienced:

Stroke Length, X_0

$$X_0 = \frac{\sqrt{R}\left(\frac{V_{L2}}{A_2} + S\right) - \left(\frac{V_{L1}}{A_1}\right)}{1 + \sqrt{R}}$$

$$X_0 = \frac{\sqrt{1.8}\left(\frac{13.4}{1} + 25.6\right) - \left(\frac{1}{1.8}\right)}{1 + \sqrt{1.8}}$$

$$X_0 = \frac{1.342(13.4 + 25.6) - .556}{1 + 1.342}$$

$$X_0 = 22.1 \; in$$

7. DETERMINE LOAD STIFFNESS

Find the hydraulic stiffness of the system:

From step #2:

$$A_1 = 1.8 \; in^2$$
$$A_2 = 1.0 \; in^2$$

Full Bore Volume at X_0

$$V_1 = A_1 \times X_0$$
$$V_1 = 1.8 \; in^2 \times 22.1 \; in$$
$$V_1 = 39.8 \; in^3$$

Rod End Volume at X_0

$$V_2 = A_2 \times (S - X_0)$$
$$V_2 = 1 \; in^2 \times (25.6 \; in - 22.1 \; in)$$
$$V_2 = 3.5 \; in^3$$

Hydraulic Stiffness, C_H

$$C_H = \frac{(E)A_1}{V_{L1} + V_1} + \frac{(E)A_2}{V_{L2} + V_2}$$

$$C_H = \frac{2 \times 10^5 (1.8)^2}{1.0 + 39.8} + \frac{2 \times 10^5 (1.0)^2}{13.4 + 3.5}$$

$$C_H = 2.8 \times 10^4 \ lb_f/in$$

8. DETERMINE LOAD NATURAL FREQUENCY
Find the natural frequency of the load:

Load Mass, M

$$W = M \times g$$

$$M = \frac{W}{g}$$

$$M = \frac{W}{32.2 \ ft/sec^2 \times 12 \ in/ft}$$

$$M = \frac{660 \ lb_f}{386 \ in/sec^2}$$

$$M = 1.71 \ lb_f \, sec^2/in$$

Load Natural Frequency, ω_L

$$\omega_L = \sqrt{\frac{C_H}{M}}$$

$$\omega_L = \sqrt{\frac{2.8 \times 10^4 \ lb_f/in}{1.71 \ lb_f \, sec^2/in}}$$

$$\omega_L = 128 \ radians/sec \quad (or \ 20.4 \ Hz)$$

---

9. DETERMINE VALVE NATURAL FREQUENCY

Find the natural frequency of the valve:

From APPENDIX E – Catalog Data (page 279)

In step #5, we chose an SM4 – 10(5)19 valve.

Refer to graph of frequency response for the SM4–10/15/20 valves at 19 and 28 l/min (5.0 and 7.5 US gpm).

> The two downward curving lines indicate amplitude ratios.
> The two upward curving lines indicate phase lags.

> The solid line curves show response at 40% current.
> The dotted line curves show response at full 100% current.

> Refer to the upward curving, dotted line. This curve crosses the 90° phase lag line at a frequency of about 85 Hz.

Therefore, the rated natural frequency of this valve (at 90° phase shift and 3000 PSI system pressure) is

$$85 \, Hz$$

and since

$$\omega_V = 2\pi f_V$$

then

$$\omega_V = 2 \times 3.1416 \times 85 \, cycles/sec$$
$$\omega_V = 534 \, radians/sec \quad (or \, 85 \, Hz)$$

Note that this value is valid only with a system pressure of 3000 PSI.

Since our actual system pressure is only 1740 PSI, we must apply a **correction factor** to this value of valve natural frequency.

From APPENDIX E – Catalog Data (page 281)

Refer to the graph in the middle left column, "Changes in Frequency Response with Pressure."

> P is our system pressure (1740 PSI)
> P_S is rated pressure (3000 PSI)

Therefore,

$$\frac{P}{P_S} = \frac{1740}{3000}$$
$$\frac{P}{P_S} = .6$$

Observe where the vertical $P/P_S = .6$ line (labeled on the bottom of the graph) crosses the response line, and read the corresponding value of f_P/f_{P_S}.

$$\text{CORRECTION FACTOR } \frac{f_P}{f_{P_S}} \text{ is about } 0.84.$$

(From the explanation given next to the graph, you can see that

f_P is the valve frequency at system pressure and
f_{P_S} is the valve frequency at 3000 PSI)

Therefore,

Valve Natural Frequency, ω_V

ω_V = *Frequency at 90° Phase*   x   *Correction Factor for System*
       Shift and P_S = 3000 PSI   *Pressure of 1740 PSI*

ω_V = 534 *radians/sec* × 0.84
ω_V = 449 *radians/sec* (*or 71.4 Hz*)

10. DETERMINE SYSTEM NATURAL FREQUENCY (ω_S)

From Step #8,

$$\omega_L = 128 \; radians/sec$$

therefore

$$3\omega_L = 384 \; radians/sec$$

From Step #9,

$$\omega_V = 449 \; radians/sec$$
$$\text{Since, } \omega_V > 3\omega_L$$
$$\text{then } \omega_S = \omega_L$$
$$\omega_S = 128 \; radians/sec$$

11. DETERMINE OPEN LOOP GAIN (K_{VX})

$$K_{VX} = 0.2\omega_S$$
$$K_{VX} = 0.2 \times 128 \; radians/sec$$
$$K_{VX} = 25.6 \; sec^{-1}$$

12. DETERMINE RATED FLOW AT SYSTEM PRESSURE (Q_{RP})

Convert Q_R from GPM to in^3/sec

$$Q_R = 5\frac{gallons}{minute} \times 231\frac{in^3}{gallon} \times \frac{1\,min}{60\,sec}$$
$$Q_R = 19.25\frac{in^3}{sec}$$

Calculate Q_{RP}

$$Q_{RP} = Q_R\sqrt{\frac{P_S}{1000}}$$
$$Q_{RP} = 19.25\sqrt{\frac{1740}{1000}}$$
$$Q_{RP} = 25.4 \; in^3/sec$$

13. DETERMINE TOTAL POSITION ERROR (ΔX_{TOT})

$$\Delta X_{TOT} = \Delta X_U + \Delta X_E + \Delta X_{FB} + \Delta X_{BL}$$

From the Specifications:

$$\Delta X_E = 0$$
$$\Delta X_{FB} = 0$$
$$\Delta X_{BL} = 0$$

and therefore

$$\Delta X_{TOT} = \Delta X_U$$

Calculate ΔX_U

$$\Delta X_U = 0.04 \frac{Q_{RP}}{K_{vx} A_2}$$

$$\Delta X_U = 0.04 \frac{25.4 \ in^3/sec}{25.6 \ sec^{-1} \times 1.0 in^2}$$

$$\Delta X_U = 0.0397 \ in$$

Since the above calculations tend to produce a conservative estimate of the positional error, it can be assumed that the total error will not exceed 0.04 inches.

The specification can be met.

Example 8.2:

Figure 8.2 shows a hydraulic test rig, used to test landing gear for fighter planes.

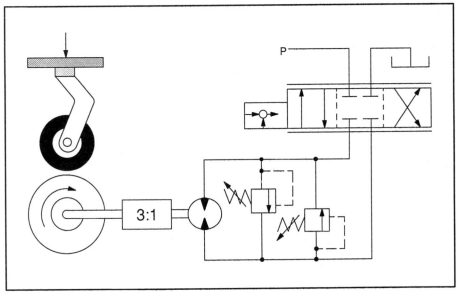

Figure 8.2

A hydraulic motor turns a drum via a speed increasing gearbox. The drum is accelerated up to maximum speed, and then the landing gear is lowered onto the drum, and the wheel brakes on the landing gear are applied. The drum is then gradually decelerated to a stop by applying a ramp signal to the servo valve amplifier.

DETERMINE:

1. The size of the servo valve required for this application, and
2. The gain setting that will be required in the amplifier.

ENGINEERING SPECIFICATIONS:

Effective Inertia of Drum . . . . 6.2 in lb_f sec^2

Maximum Drum Speed . . . . 420 RPM

Load Torque . . . . 317 lb_f ft

Motor Displacement . . . . 45 in^3/rev

Motor Damping Torque . . . . 37 lb_f ft

Motor Volumetric Efficiency (ε_{vol}) 96%

Pipe volume per side (servo valve to motor) . . . . 4.9 in^3

System Pressure . . . . 2800 PSI

Return (tank) line pressure . . . . 29 PSI

Gearbox ratio (Motor:Drum) 1:3

Gearbox Efficiency 90%

Maximum Acceleration Time (t_{max}) . . . . 5 seconds

Tachometer-Generator Gain (H_{FB}) . . . . 0.071 V/RPM

1. DETERMINE MOTOR TORQUE REQUIRED TO ACCELERATE
    LOAD
    Find the torque necessary to bring the drum up to maximum speed:

Acceleration Torque = Effective x Angular + Damping
                      Inertia   Acceleration Torque

$$T_a = (J_{EFF} \times \alpha) + T_D$$

Effective Inertia, J_{EFF}

$$J_{EFF} = J \times n^2$$
$$n = \frac{output}{input} = 3 \quad (Gearbox\ ratio)$$
$$J_{EFF} = 6.2\ in\ lb_f sec^2 \times 3^2$$
$$J_{EFF} = 55.8\ in\ lb_f\ sec^2$$

Angular Acceleration, α

Maximum drum speed is 420 RPM

$$Maximum\ motor\ speed = \frac{420\ RPM}{n}$$
$$Maximum\ motor\ speed = \frac{420\ RPM}{3}$$
$$Maximum\ motor\ speed = 140\ RPM$$

$$\omega_{motor} = 140 \frac{revolutions}{minute} \times 2\pi \frac{radians}{revolution} \times \frac{1\ minute}{60\ seconds}$$
$$\omega_{motor} = 140 \times 2\pi \times \frac{1}{60}$$
$$\omega_{motor} = 14.7\ radians/second$$

Maximum Acceleration Time is t_{accel} = 5 seconds

Therefore, angular acceleration is

$$\alpha = \frac{\omega_{motor}}{t_{accel}}$$

$$\alpha = \frac{14.7 \ radians/second}{5 \ seconds}$$

$$\alpha = 2.94 \ radians/second^2$$

Damping Torque, T_D

$$T_D = 37 \ lb_f ft \times 12 \ \frac{in}{ft}$$

$$T_D = 444 \ lb_f in$$

Acceleration Torque, T_a

$$T_a = (J_{EFF} \times \alpha) + T_D$$

$$T_a = (55.8 \times 2.94) + (444)$$

$$T_a = 608 \ lb_f in$$

T_a would be 608 lb_f in, IF the gearbox was 100% efficient. However, it is only 90% efficient.

Therefore:

$$.90 \times T_a = 608 \ lb_f in$$

$$T_a = \frac{608 \ lb_f in}{.90}$$

$$T_a = 676 \ lb_f in$$

2. DETERMINE MOTOR TORQUE REQUIRED TO DRIVE LOAD
    Find the torque required to drive the landing gear:

Motor Torque = Load Torque x Gearbox ratio + Damping Torque

$$T_M = (T_E \times n) + T_D$$

Load Torque, T_E

$$T_E = 317 \, lb_f ft \times 12 \frac{in}{ft}$$
$$T_E = 3,804 \, lb_f in$$

Motor Torque, T_M

$$T_M = (T_E \times n) + T_D$$
$$T_M = (3,804 \times 3) + 444 \, lb_f in$$
$$T_M = 11,856 \, lb_f in$$

T_M would be 11,856 lbf in, IF the gearbox was 100% efficient. However, it is only 90% efficient.

Therefore:

$$.90 \times T_M = 11,856 \, lb_f in$$
$$T_M = \frac{11,856 \, lb_f in}{.90}$$
$$T_M = 13,173 \, lb_f in$$

3. **DETERMINE ACTUATOR PRESSURES P_1 & P_2**
   Find the pressures at the other two valve ports:

Pressure, P_1

$$P_1 = \frac{P_S + P_T}{2} + \frac{\pi T}{D_M}$$

$$P_1 = \frac{2,800 + 29}{2} + \frac{3.1416 \times 13,173}{45}$$

$$P_1 = 2,334 \, PSI$$

Pressure, P_2

$$P_2 = P_S - P_1 + P_T$$

$$P_2 = 2,800 - 2,334 + 29 \, PSI$$

$$P_2 = 495 \, PSI$$

4. **DETERMINE ACTUAL FLOW**
   Find the actual flow required to the motor:

Actual Flow, Q_A

$$Q_A = \frac{RPM \times D_M}{231 \times \varepsilon_{vol}}$$

$$Q_A = \frac{140 \, rev/min \times 45 \, in^3/rev}{231 \, in^3/gal \times .96}$$

$$Q_A = 28.4 \, GPM$$

5. DETERMINE RATED FLOW OF VALVE

Find the minimum flow rating of the valve for this application:

Rated Flow, Q_R

$$Q_R = Q_A \sqrt{\frac{500}{P_S - P_1}}$$

$$Q_R = 28.4 \times \sqrt{\frac{500}{2,800 - 2,334}}$$

$$Q_R = 28.4 \times \sqrt{1.073}$$

$$Q_R = 29.4 \; GPM$$

Minimum rated flow of valve to be used for this application is therefore: 29.4 GPM

The smallest valve in Appendix E (page 303, Model Code info) meeting this spec is: SM4 – 40(30)113 rated for 30 GPM (113 l/min)

6. DETERMINE ACTUATOR STIFFNESS

Find the hydraulic stiffness of the actuator:

Hydraulic Stiffness, C_H

From the specifications,

$$V_{L1} = V_{L2} = 4.9 \; in^3$$

$$C_H = E \times \left(\frac{D_M}{2\pi}\right)^2 \times \frac{2}{V_{L1} + \frac{D_M}{2}}$$

$$C_H = 2 \times 10^5 \left(\frac{45}{2\pi}\right)^2 \times \frac{2}{4.9 + \frac{45}{2}}$$

$$C_H = 200,000 \times 51.294 \times 0.073$$

$$C_H = 7.5 \times 10^5 \; lb_f \, in/radian$$

7. DETERMINE ACTUATOR NATURAL FREQUENCY

    Find the natural frequency of the actuator (Assume all connections from motor to drum are rigid and have infinite stiffness):

Natural Frequency, ω_L

$$\omega_L = \sqrt{\frac{C_H}{J_{EFF}}}$$

$$\omega_L = \sqrt{\frac{7.5 \times 10^5}{55.8}}$$

$$\omega_L = 116\ radians/sec \quad (18.5\ Hz)$$

8. DETERMINE VALVE NATURAL FREQUENCY

    Find the natural frequency of an SM4–40(30)113 valve.

From APPENDIX E – Catalog Data (page 280)

In step #5, we chose an SM4 – 40(30)113 valve.

Refer to graph of frequency response for the SM4–40 valves at 95 and 113 l/min (25 and 30 USgpm).

The upward curving line crosses the 90° phase lag line at a frequency of about 42 Hz.

Therefore, the rated natural frequency of this valve (at 90° phase shift and 3000 PSI system pressure) is

> 42 Hz

and since

$$\omega_V = 2\pi f_V$$

then

$$\omega_V = 2 \times 3.1416 \times 42\ cycles/sec$$
$$\omega_V = 264\ radians/sec \quad (42\ Hz)$$

Note that this value is valid only with a system pressure of 3000 PSI.

Since our actual system pressure is only 2,800 PSI, we must apply a **correction factor** to this value of valve natural frequency.

From APPENDIX E – Catalog Data (page 281)

Refer to the graph in the middle left column, "Changes in Frequency Response with Pressure."

P is our system pressure (2,800 PSI)

P_S is rated pressure (3,000 PSI)

Therefore,

$$\frac{P}{P_S} = \frac{2,800}{3,000}$$

$$\frac{P}{P_S} = .93$$

Estimate where the vertical $P/P_S = .93$ line (labeled on the bottom of the graph) crosses the response line, and read the corresponding value of f_P/f_{P_S}.

$$\text{CORRECTION FACTOR } \frac{f_P}{f_{P_S}} \text{ is about } 0.97$$

(From the explanation given next to the graph, you can see that

f_P is the valve frequency at system pressure and

f_{P_S} is the valve frequency at 3000 PSI)

Therefore,

Valve Natural Frequency, ω_V

$$\omega_V = \begin{array}{c} \textit{Frequency at } 90° \\ \textit{Phase Shift} \\ \textit{and } P_S = 3,000 \textit{ PSI} \end{array} \quad x \quad \begin{array}{c} \textit{Correction Factor} \\ \textit{for System Pressure} \\ \textit{of 2,800 PSI} \end{array}$$

$$\omega_V = 264 \textit{ radians/sec} \times 0.97$$

$$\omega_V = 256 \textit{ radians/sec} \quad (40.7 \textit{ Hz})$$

9. DETERMINE SYSTEM NATURAL FREQUENCY

Find the natural frequency of the overall system:

$$\omega_V = 256 \ radians/second \text{ and}$$
$$\omega_L = 116 \ radians/second \text{ so}$$
$$3\omega_L = 348 \ radians/second$$

Since ω_v is less than $3\omega_L$,

$$\omega_S = \frac{\omega_L \times \omega_V}{\omega_L + \omega_V}$$
$$\omega_S = \frac{116 \times 256}{116 + 256}$$
$$\omega_S = 79.8 \ radians/second$$

10. DETERMINE OPEN LOOP GAIN

Find K_{vw} for this system:

$$K_{VW} = 0.2 \times \omega_S$$
$$K_{VW} = 0.2 \times 79.8 \ radians/sec$$
$$K_{VW} = 16 \ radians/sec$$

11. DETERMINE AMPLIFIER GAIN

Calculate the proper gain setting for the valve's amplifier:

$$K_{VW} = \frac{G_{AMP} \times G_{SV} \times H_{FB}}{\frac{D_M}{2\pi}}$$

$$G_{AMP} = \frac{K_{VW} \times \frac{D_M}{2\pi}}{G_{SV} \times H_{FB}}$$

Servo Valve Gain, G_{SV}

$$G_{SV} = \frac{output}{input} = \frac{Load\ flow\ (in^3/sec)}{Rated\ drive\ current\ (ma)}$$

We must now further refine our choice of SM4 servo valve for this calculation. For this example, we will choose the

SM4 – 40(30)113 – 20/200 – 10,

which has a 20 ohm, 200 ma rated torque motor coil. Therefore,

$$G_{SV} = \frac{Q_A}{200\ ma}$$

From step #4,

$$Q_A = 28.4\ GPM$$

Convert to in^3/sec:

$$Q_A = 28.4\frac{gallons}{minute} \times 231\frac{in^3}{gallon} \times \frac{1\ minute}{60\ seconds}$$

$$Q_A = 109.3\frac{in^3}{sec}$$

Therefore,

$$G_{SV} = \frac{Q_A}{200\ ma}$$

$$G_{SV} = \frac{109.3\ in^3/sec}{200\ ma}$$

$$G_{SV} = 0.55\frac{in^3/sec}{ma}$$

Feedback Transducer Gain, H_{FB}

$$H_{FB} \;=\; 0.071 \; V/RPM$$

$$H_{FB} \;=\; 0.071 \frac{V}{rev/min} \times \frac{1 \; rev}{2\pi \; rad} \times \frac{60 \; sec}{1 \; min}$$

$$H_{FB} \;=\; 0.68 \frac{V}{rad/sec}$$

Amplifier Gain, G_{amp}

$$G_{AMP} \;=\; \frac{K_{VW} \times \frac{D_M}{2\pi}}{G_{SV} \times H_{FB}}$$

$$G_{AMP} \;=\; \frac{16 \times \frac{45}{2\pi}}{0.55 \times 0.68}$$

$$G_{AMP} \;=\; 306.4 \frac{ma/sec}{volt}$$

Example 8.3:

Figure 8.3 shows the system used to form rivets in a riveting machine.

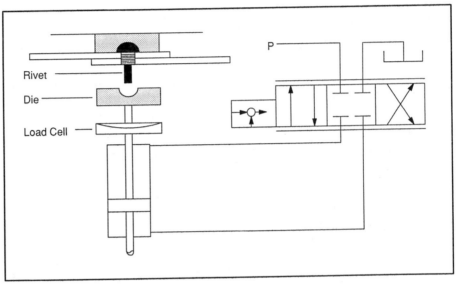

Figure 8.3

A double rod cylinder presses the die onto the rivet. Pressing force is sensed by the load cell, which is placed between the cylinder rod and the moving die.

In order to assure uniform rivet joint formation, the specification calls for the pressing force on the rivet to be controlled to an accuracy of 1% over the working pressure range.

Determine whether this can be achieved with a standard proportional type amplifier, or whether a more expensive proportional plus integral (P+I) will be required.

ENGINEERING SPECIFICATIONS:

Cylinder Bore Diameter . . . . 8.0 in
Cylinder Rod Diameter . . . . 4.5 in
Cylinder Stroke . . . . 5.0 in
Cylinder Seal Friction . . . . 100 lb_f
Load Weight (Die plus load cell plus rod) . . . . 400 lb_f
Pressing Force (minimum) . . . . 35,000 lb_f
Pressing Force (maximum) . . . . 70,000 lb_f
Pressing Force Range is . . . . 35,000 lb_f
1% of Force Range is . . . . 350 lb_f
Opening/Closing Velocity . . . . 2.0 in/sec
Velocity During Press Operation . . . . 0.04 in/sec
Maximum Supply Pressure . . . . 2,500 PSI
Return (Tank) Line Pressure . . . . 50 PSI
Bulk Modulus of Hydraulic Fluid . . . . 2×10^5 lb_f/in^2
Pipe Volume, Valve to Cylinder (each side) . . . . 2.5 in^3
Mechanical Stiffness of System . . . . infinite
Load Cell Accuracy . . . . .1% of range

1. DETERMINE MAXIMUM LOAD PRESSURE
   Find the maximum pressure required to generate the maximum specified pressing force.

Piston Area, A

$$A_1 = A_2 = A \quad (Double\ Rod\ Cylinder)$$
$$A = \frac{\pi \times 8^2}{4} - \frac{\pi \times 4.5^2}{4}$$
$$A = 34.4\ in^2$$

Load Pressure, P_L

Maximum Pressing Force (F_P) is 70,000 lb_f
Weight of Load to Raise (W) is 400 lb_f
Backpressure Force in Cylinder is

$$F_{BP} = P_T \times A$$
$$F_{BP} = 50\ lb_f/in^2 \times 34.4\ in^2$$
$$F_{BP} = 1,720\ lb_f$$

Therefore, by force balancing,

$$P_L = \frac{F_P + W + F_{BP}}{A}$$
$$P_L = \frac{70,000 + 400 + 1,720\ lb_f}{34.4\ in^2}$$
$$P_L = 2,097\ PSI$$

2. DETERMINE P_1 AND P_2 DURING RAPID TRAVERSE PORTION OF CYCLE
   Find P_1 and P_2 during raising of the cylinder, before it is actually applying force to the rivet.

Total Force, F (Raising)
F = Acceleration Force + Cylinder Seal Force + Load Weight + Load Friction

$$F = F_a + F_s + W + F_E$$

Since the load is rubbing against air only, assume $F_E = 0$.
Since the load is only 400 lb_f and is not accelerated rapidly, assume $F_a = 0$.

From the specification, $F_s = 100$ lb$_f$ and $W = 400$ lb$_f$.
Therefore,

$$F = 0 + 100 + 400 + 0 \ lb_f$$

$$F = 500 \ lb_f$$

Bottom Pressure, P_1 (Raising)

$$P_1 = \frac{P_s A + F + P_T A}{2A}$$

$$P_1 = \frac{2500 \times 34.4 + 500 + 50 \times 34.4}{2 \times 34.4}$$

$$P_1 = 1,282 \ PSI$$

Top Pressure, P_2 (Raising)

$$P_2 = P_T + \frac{P_s - P_1}{R^2} \qquad \text{(Note: } A_1 = A_2 \text{, so } R = 1\text{)}$$

$$P_2 = 50 + \frac{2500 - 1282}{1^2}$$

$$P_2 = 1,268 \ PSI$$

Find P_1 and P_2 during lowering of cylinder, after it has actually applied force to the rivet.

Total Force, F (Lowering)

F = Acceleration Force + Cylinder Seal Force − Load Weight + Load Friction

$$F = F_a + F_s - W + F_E$$

Since the load is rubbing against air only, assume $F_E = 0$.

Since the load is only 400 lbf and is not accelerated rapidly, assume $F_a = 0$.

From the specification, Fs = 100 lb$_f$, and W = 400 lb$_f$. Therefore,

$$F = 0 + 100 - 400 + 0 \ lb_f$$

$$F = -300 \ lb_f$$

Top Pressure, P_2 (Lowering)

$$P_2 = \frac{P_S A R^3 + F + P_T A R}{A(1 + R^3)}$$

$$P_2 = \frac{2500 \times 34.4 \times 1^3 - 300 + 50 \times 34.4 \times 1}{34.4(1 + 1^3)}$$

$$P_2 = 1,271 \ PSI$$

Bottom Pressure, P_1 (Lowering)

$$
\begin{aligned}
P_1 &= P_T + R^2(P_s - P_2) \\
P_1 &= 50 + 1^2(2,500 - 1,271) \\
P_1 &= 1,279 \; PSI
\end{aligned}
$$

3. DETERMINE LOAD FLOW

Find the actual flow required to move the actuator at maximum velocity as specified:

Load Flow, Q_A

$$
\begin{aligned}
Q_A &= V_{max} \times A \\
Q_A &= 2 \; in/sec \times 34.4 \; in^2 \\
Q_A &= 68.8 \; in^3/sec \\
Q_A &= 68.8 \; in^3/sec \times \frac{1 \; gallon}{231 \; in^3} \times \frac{60 \; seconds}{1 \; minute} \\
Q_A &= 17.9 \; GPM
\end{aligned}
$$

4. DETERMINE VALVE RATED FLOW

Select a servo valve properly sized for this application:

Rated Flow, Q_R (Raising)

$$
\begin{aligned}
Q_R &= Q_A \sqrt{\frac{500}{P_S - P_1}} \\
Q_R &= 17.9 \sqrt{\frac{500}{2,500 - 1,282}} \\
Q_R &= 17.9 \sqrt{\frac{500}{1,218}} \\
Q_R &= 11.5 \; GPM
\end{aligned}
$$

Rated Flow, Q_R (Lowering)

$$
\begin{aligned}
Q_R &= Q_A \sqrt{\frac{500}{P_S - P_2}} \\
Q_R &= 17.9 \sqrt{\frac{500}{2,500 - 1,271}}
\end{aligned}
$$

$$Q_R = 17.9 \sqrt{\frac{500}{1,229}}$$

$$Q_R = 11.4 \; GPM$$

Minimum rated flow of valve to be used for this application is therefore:

$$11.5 \; GPM$$

**The smallest valve in Appendix E (page 302, Model Code info)
meeting this spec is: SM4 – 15(12.5)47 rated for 12.5 GPM (47 l/min)**

5. DETERMINE LOAD STIFFNESS

Find the hydraulic stiffness of the system (Mechanical stiffness is specified
as "infinite" and can therefore be ignored):

Hydraulic Stiffness, C_H

$$C_H = EA^2 \left(\frac{1}{V_{L1} + \frac{S}{2} \times A} + \frac{1}{V_{L2} + \frac{S}{2} \times A} \right)$$

Since $V_{L1} = V_{L2} = V_L$

$$C_H = EA^2 \times 2 \left(\frac{1}{V_L + \frac{S}{2} \times A} \right)$$

$$C_H = 2 \times 10^5 \times 34.4^2 \times 2 \left(\frac{1}{2.5 + \frac{S}{2} \times 34.4} \right)$$

$$C_H = 5.35 \times 10^6 \; lb_f / in$$

6. DETERMINE LOAD NATURAL FREQUENCY
   Find the natural frequency of the actuator/load combination:

Load Mass, M

$$M = \frac{W}{g}$$

$$g = 32.2\,ft/sec^2 \times 12\frac{in}{ft} = 386.4\,in/sec^2$$

$$M = \frac{400\,lb_f}{386.4\,in/sec^2}$$

$$M = 1.035\,lb_f\,sec^2/in$$

Load Natural Frequency, ω_L

$$\omega_L = \sqrt{\frac{C_H}{M}}$$

$$\omega = \sqrt{\frac{5.35 \times 10^6}{1.035}}$$

$$\omega_L = 2274\,radians/second \quad (360\,Hz)$$

7. DETERMINE VALVE NATURAL FREQUENCY

Find the natural frequency of an SM4–15(12.5)47 valve.

From APPENDIX E – Catalog Data (page 280)

In step #4, we chose an SM4 – 15(12.5)47 valve.

Refer to graph of frequency response for the SM4–15 valves at 47 and 57 l/min (12.5 and 15 USgpm). The upward curving dotted line crosses the 90° phase lag line at a frequency of about 60 Hz.

Therefore, the rated natural frequency of this valve (at 90° phase shift and 3000 PSI system pressure) is

$$60 \ Hz$$

Since

$$\omega_V \ = \ 2\pi f_V$$

then

$$\omega_V \ = \ 2 \times 3.1416 \times 60 \ cycles/sec$$
$$\omega_V \ = \ 377 \ radians/sec \quad (60 \ Hz)$$

Note that this value is valid only with a system pressure of 3000 PSI.

Since our actual system pressure is only 2,500 PSI, we must apply a **correction factor** to this value of valve natural frequency.

From APPENDIX E – Catalog Data (page 281)

Refer to the graph in the middle left column, "Changes in Frequency Response with Pressure."

> P is our system pressure (2,500 PSI)
> P_S is rated pressure (3,000 PSI)

Therefore,

$$\frac{P}{P_S} \ = \ \frac{2,500}{3,000}$$
$$\frac{P}{P_S} \ = \ .83$$

Estimate where the vertical P/P$_S$ =.83 line (labeled on the bottom of the graph) crosses the response line, and read the corresponding value of f$_P$/f$_{Ps}$.

$$\text{CORRECTION FACTOR } \frac{f_P}{f_{P_s}} \text{ is about } 0.94$$

(From the explanation given next to the graph, you can see that

f_P is the valve frequency at system pressure and
f_{Ps} is the valve frequency at 3000 PSI)

Therefore,

Valve Natural Frequency, ω_V

ω_V = *Frequency at 90° Phase Shift x Correction Factor for System Pressure*
and P_s = 3,000 PSI   *of 2,500 PSI*

$$\omega_V = 377 \; radians/sec \times 0.94$$
$$\omega_V = 354 \; radians/sec \quad (56.4 \; Hz)$$

8. DETERMINE SYSTEM NATURAL FREQUENCY
Find the natural frequency of the overall system:

Natural Frequency, ω_L

$$\omega_V = 354 \; radians/sec \text{ so,}$$
$$3\omega_V = 1062 \; radians/sec \text{ and,}$$
$$\omega_L = 2274 \; radians/sec.$$

Therefore, $\omega_L > 3\omega_v$ indicating that

$$\omega_S = \omega_V$$
$$\omega_S = 354 \; radians/sec$$

9. DETERMINE OPEN LOOP GAIN
Find K_{VP} for this system:

Open Loop Gain, K_{VP}

$$K_{VP} = 0.4\omega_S$$
$$K_{VP} = 0.4 \times 354$$
$$K_{VP} = 141.6 \; sec^{-1}$$

10. DETERMINE RATED FLOW AT SYSTEM PRESSURE

This servo valve is rated for 12.5 GPM at 3,000 PSI. Find its rated flow at the stated system pressure of 2,500 PSI.

Rated Flow at 3000 PSI, Q_R

$$Q_R = 12.5 \frac{gallons}{minute} \times 231 \frac{in^3}{gallon} \times \frac{1 \ minute}{60 \ seconds}$$

$$Q_R = 48.1 \ in^3/sec$$

Rated Flow at 2500 PSI, Q_{RP}

$$Q_{RP} = Q_R \sqrt{\frac{P_s}{1000}}$$

$$Q_{RP} = 48.1 \sqrt{\frac{2500}{1000}}$$

$$Q_{RP} = 76.1 \ in^3/sec$$

11. DETERMINE PRESSURE ERROR DUE TO VALVE UNCERTAINTIES

Find the pressure error due to valve factors other than internal leakage:

Pressure Error, ΔP_U

$$\Delta P_U = 4 \times 10^{-2} \frac{C_H \times Q_{RP}}{A^2 \times K_{VP}}$$

$$\Delta P_U = 4 \times 10^{-2} \times \frac{5.35 \times 10^6 \times 76.1}{34.4^2 \times 141.6}$$

$$\Delta P_U = 97.2 \ PSI$$

12. DETERMINE PRESSURE ERROR DUE TO LEAKAGE

Find the pressure error produced by internal leakage across the valve:

Pressure Error, ΔP_L

$$\Delta P_L = 2 \times 10^{-2} \left(\frac{C_H \times Q_{RP}}{A^2 \times K_{VP}} \right) \left(\frac{\Delta P_{AB}}{P_S} \right)$$

We are attempting to control the pressing force between 35,000 lb$_f$ and 70,000 lb$_f$. We must therefore find the value of ΔP_{AB} which is the higher for these two extremes.

Pressure Drop Across Valve at 35,000 lb$_f$, ΔP_{AB}

$$R = 1$$
$$A = 34.4 in^2$$
$$P_S = 2,500\, PSI$$
$$P_T = 50\, PSI$$

$$F = F_a + F_c + F_E + F_S + W$$
$$F = 0 + 35,000 + 0 + 100 + 400$$
$$F = 35,500\, lb_f$$

Total minimum force is 35,500 lb.

$$P_1 = \frac{P_S A + R^2(F + P_T A)}{A(1 + R^3)}$$
$$P_1 = \frac{2,500(34.4) + 1^2(35,500 + 50(34.4))}{34.4(1 + 1^3)}$$
$$P_1 = 1,791\, PSI$$

$$P_2 = P_T + \frac{P_S - P_1}{R^2}$$
$$P_2 = 50 + \frac{2,500 - 1,791}{1^2}$$
$$P_2 = 759\, PSI$$

$$\Delta P_{AB} = P_1 - P_2$$
$$\Delta P_{AB} = 1,791 - 759\, PSI$$
$$\Delta P_{AB} = 1,032\, PSI \quad (\,at\, 35,000\, lb_f)$$

Pressure Drop Across Valve at 70,000 lb$_f$, ΔP_{AB}

$$R = 1$$
$$A = 34.4\ in^2$$
$$P_S = 2,500\ PSI$$
$$P_T = 50\ PSI$$

$$F = F_a + F_c + F_E + F_S + W$$
$$F = 0 + 70,000 + 0 + 100 + 400$$
$$F = 70,500\ lb_f$$

Total minimum force is 70,500 lb.

$$P_1 = \frac{P_S A + R^2(F + P_T A)}{A(1 + R^3)}$$
$$P_1 = \frac{2,500(34.4) + 1^2(70,500 + 50(34.4))}{34.4(1 + 1^3)}$$
$$P_1 = 2,300\ PSI$$

$$P_2 = P_T + \frac{P_S - P_1}{R^2}$$
$$P_2 = 50 + \frac{2,500 - 2,300}{1^2}$$
$$P_2 = 250\ PSI$$

$$\Delta P_{AB} = P_1 - P_2$$
$$\Delta P_{AB} = 2,300 - 250\ PSI$$
$$\Delta P_{AB} = 2,050\ PSI \quad (at\ 70,000\ lb_f)$$

Of the two force extremes, the value of ΔP_{AB} at 70,000 lb$_f$ is the higher value. Therefore, use:

$$\Delta P_{AB} = 2,050\ PSI$$

Pressure Error, ΔP_L

$$\Delta P_L = 2 \times 10^{-2}\left(\frac{C_H \times Q_{RP}}{A^2 \times K_{VP}}\right)\left(\frac{\Delta P_{AB}}{P_S}\right)$$
$$\Delta P_L = \frac{2 \times 10^{-2} \times 5.35 \times 10^6 \times 76.1 \times 2,050}{34.4^2 \times 141.6 \times 2,500}$$

$$\Delta P_L \;=\; 39.8\,PSI$$

13. DETERMINE PRESSURE ERROR DUE TO ACTUATOR MOTION
Find the pressure error caused by the speed of the actuator:

Pressure Error, ΔP_V

$$\Delta P_V \;=\; \frac{V \times C_H}{A \times K_{VP}}$$
$$\Delta P_V \;=\; \frac{0.04 \times 5.35 \times 10^6}{34.4 \times 141.6}$$
$$\Delta P_L \;=\; 43.9\,PSI$$

14. DETERMINE TOTAL PRESSURE ERROR
Find the total error caused by all factors combined:

Total Error, ΔP_{TOT}

$$\Delta P_{TOT} \;=\; \Delta P_U + \Delta P_L + \Delta P_V + \Delta P_{FB}$$

ΔP_{FB} is .1% of range (use 2,500 PSI)

$$\Delta P_{TOT} \;=\; 97.2 + 39.8 + 43.9 + .001(2,500)$$
$$\Delta P_{TOT} \;=\; 183.4\,PSI$$

15. DETERMINE TOTAL FORCE ERROR

Determine whether the specification can be met with a simple proportional amplifier:

The total pressure error is 183.4 PSI. This represents a force error of

$$183.4 \ lb_f/in^2 \times 34.4 \ in^2 \ = \ 6,309 \ lb_f$$

At the 35,000 lb$_f$ minimum, this represents a percentage error of

$$\frac{6,309}{35,000} \times 100\% \ = \ 18.0\%$$

At the 70,000 lb$_f$ maximum, this represents a percentage error of

$$\frac{6,309}{70,000} \times 100\% \ = \ 9.0\%$$

Since the specification requires that there be no more than 1% error, it is highly unlikely that the spec can be met using a proportional type control amplifier.

This system will need a P+I type amplifier for the valve, which will reduce the error to virtually zero when properly adjusted.

Linear System Relationships

Linear Systems

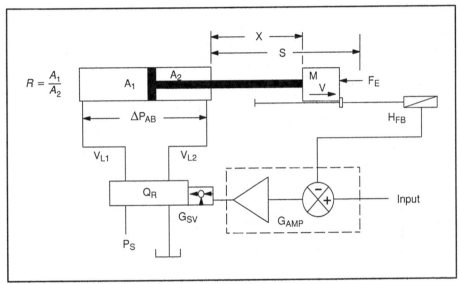

Figure A.1

1. Determine Cylinder Area Ratio (R)

$$A_1 \rightarrow \boxed{R = \frac{A_1}{A_2}} \leftarrow A_2$$

$$\downarrow R$$

2. Determine Stroke Length For Minimum Stiffness (X_m)

$$
\begin{array}{c}
A_1 \rightarrow \\
A_2 \rightarrow \\
V_{L1} \rightarrow \\
V_{L2} \rightarrow \\
S \rightarrow
\end{array}
\boxed{X_m = \frac{\sqrt{R}\left(\frac{V_{L2}}{A_2} + S\right) - \frac{V_{L1}}{A_1}}{1 + \sqrt{R}}}
$$

$$\downarrow X_m$$

3. Determine Hydraulic Stiffness (C_H)

X_m

A_1
A_2
V_{L1}
V_{L2}
S
E

$$C_H = E \left[\frac{A_1^2}{V_{L1} + A_1 X_m} + \frac{A_2^2}{V_{L2} + A_2 (S - X_m)} \right]$$

C_H

4. Determine Total Load Stiffness (C_{TOT})

C_M

$$C_{TOT} = \frac{C_H \times C_M}{C_H + C_M}$$

C_{TOT}

5. Determine Load Natural Frequency (ω_L)

M_{EF}

$$\omega_L = \sqrt{\frac{C_{TOT}}{M_{EFF}}}$$

ω_L

6. Determine Valve Natural Frequency (ω_V)

$$\omega_V = \left(\begin{array}{c} \textit{Frequency at} \\ \textit{90° Phase Shift} \end{array} \right) \times \left(\begin{array}{c} \textit{Correction} \\ \textit{Factor For} \\ \textit{System Pressure} \end{array} \right)$$

ω_L ω_V

7. Determine System Natural Frequency (ω_s)

ω_L

ω_V

Case A: $\omega_V > 3\omega_L$

$$\omega_S = \omega_L$$

Case B: $3\omega_L > \omega_V > .3\omega_L$

$$\omega_S = \frac{\omega_L \times \omega_v}{\omega_L + \omega_v}$$

Case C: $\omega_V < .3\omega_L$

$$\omega_S = \omega_v$$

ω_S

A. POSITION CONTROL SYSTEMS:

8a. Determine Open Loop Gain of System (K_{VX})

$$\omega_S \downarrow$$

Case A:	K_{VX}	=	$0.2\omega_S$
Case B:	K_{VX}	=	$0.2\omega_S$
Case C:	K_{VX}	=	$0.4\omega_S$

$$K_{VX} \downarrow$$

9a. Determine Rated Flow at System Pressure (Q_{RP})

$$P_S \rightarrow$$
$$Q_R \rightarrow$$

$$Q_{RP} = Q_R \sqrt{\frac{P_S}{1000}}$$

$$Q_{RP} \downarrow \qquad\qquad K_{VX} \downarrow$$

10a. Determine Position Error due to Valve Uncertainties (ΔX_U)

$$A \rightarrow$$

$$\Delta X_U = 0.04 \left(\frac{Q_{RP}}{K_{VX} A} \right)$$

$$\Delta X_U \downarrow \qquad\qquad Q_{RP} \downarrow \qquad\qquad K_{VX} \downarrow$$

A. POSITION CONTROL SYSTEMS:

11a.   Determine Position Error due to External Forces (ΔX_E)

$\Delta X_U \qquad\qquad Q_{RP} \qquad\qquad K_{VX}$

A →
F_E →
P_S →

$$\Delta X_E = 0.02\left(\frac{Q_{RP}}{K_{VX}A}\right)\left(\frac{F_E}{P_S A}\right)$$

$\Delta X_U \qquad\qquad \Delta X_E \qquad\qquad K_{VX}$

12a.   Determine Total Position Error (ΔX_{TOT})

ΔX_{FB} →

$$\Delta X_{TOT} = \Delta X_U + \Delta X_E + \Delta X_{FB}$$

K_{VX}

13a.   Determine Following Error (X_F)

V →

$$X_F = \frac{V}{K_{VX}}$$

B. VELOCITY CONTROL SYSTEMS:

8b. Determine Open Loop Gain of System (K_{VV})

ω_S

Case A:	K_{VV}	=	$0.2\omega_S$
Case B:	K_{VV}	=	$0.2\omega_S$
Case C:	K_{VV}	=	$0.4\omega_S$

K_{VV}

9b. Determine Following Error (V_F)

a →

$$V_F = \frac{a}{K_{VV}}$$

C. PRESSURE CONTROL SYSTEMS:

8c. Determine Open Loop Gain of System (K_{VP})

ω_S

Case A:	K_{VP}	=	$0.2\omega_S$
Case B:	K_{VP}	=	$0.2\omega_S$
Case C:	K_{VP}	=	$0.4\omega_S$

K_{VP}

C. PRESSURE CONTROL SYSTEMS:

9c. Determine Pressure Error due to Valve Uncertainties (ΔP_U)

C_{TO} Q_{RP} K_{VP}

A →

$$\Delta P_U = 0.04\left[\frac{C_{TOT}Q_{RP}}{A^2 K_{VP}}\right]$$

ΔP_U C_{TOT} Q_{RP} K_{VP}

10c.   Determine Pressure Error due to Leakage (ΔP_L)

A →
ΔP_{AB} →
P_S →

$$\Delta P_L = 0.02\left[\frac{C_{TOT}Q_{RP}}{A^2 K_{VP}}\right]\left(\frac{\Delta P_{AB}}{P_S}\right)$$

ΔP_U C_{TOT} ΔP_L K_{VP}

11c.   Determine Pressure Error due to Actuator Motion (ΔP_V)

A →
V →

$$\Delta P_V = V\left(\frac{C_{TOT}}{K_{VP}A}\right)$$

ΔP_U ΔP_V ΔP_L K_{VP}

C. PRESSURE CONTROL SYSTEMS:

12c.   Determine Total Pressure Error (ΔP_{TOT})

ΔP_U ΔP_V ΔP_L K_{VP}

$\Delta P_{FB} \rightarrow$

$$\Delta P_{TOT} \;=\; \Delta P_U + \Delta P_L + \Delta P_V + \Delta P_{FB}$$

K_{VP}

13c.   Determine Following Error (P_F)

$\Delta P / \Delta t \rightarrow$

$$P_F \;=\; \frac{\frac{\Delta P}{\Delta t}}{K_{VP}}$$

Rotary System
Relationships

B

Rotary Systems

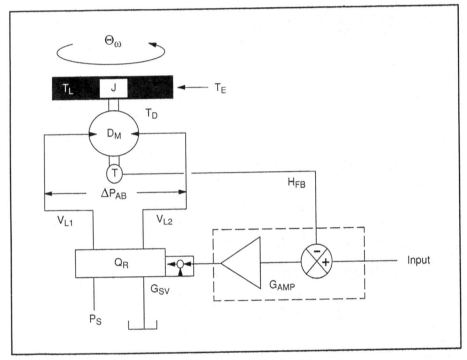

Figure B.1

1. Determine Hydraulic Stiffness (C_H)

$$D_M \rightarrow$$
$$A_1 \rightarrow$$
$$A_2 \rightarrow$$
$$V_{L1} \rightarrow$$
$$V_{L2} \rightarrow$$

$$C_H = E\frac{D_M}{2\pi}\left[\frac{A_1{}^2}{V_{L2} + \frac{D_M}{2}} + \frac{A_2{}^2}{V_{L1} + \frac{D_M}{2}}\right]$$

$$C_H \downarrow$$

2. Determine Total Stiffness (C_{TOT})

C_H
↓

C_M →

$$C_{TOT} = \frac{C_M \times C_H}{C_M + C_H}$$

C_{TOT}
↓

3. Determine Load Natural Frequency (ω_L)

C_{TOT}
↓

J_{EFF} →

$$\omega_L = \sqrt{\frac{C_{TOT}}{J_{EFF}}}$$

ω_L
↓

4. Determine Valve Natural Frequency (ω_V)

$$\omega_V = \left(\begin{array}{c} Frequency\ at \\ 90°\ Phase\ Shift \end{array} \right) \times \left(\begin{array}{c} Correction \\ Factor\ For \\ System\ Pressure \end{array} \right)$$

ω_L
↓

ω_V
↓

5. Determine System Natural Frequency (ω_s)

$$\omega_L \qquad\qquad\qquad \omega_V$$

Case A: $\omega_V > 3\omega_L$

$$\boxed{\omega_S \;=\; \omega_L}$$

Case B: $.3\omega_L < \omega_V < 3\omega_L$

$$\boxed{\omega_S \;=\; \frac{\omega_L \times \omega_V}{\omega_L + \omega_V}}$$

Case C: $\omega_V < .3\omega_L$

$$\boxed{\omega_S \;=\; \omega_V}$$

$$\omega_S$$

A. POSITION CONTROL SYSTEMS:

6a. Determine Open Loop Gain of System ($K_{V\theta}$)

$$\omega_S \downarrow$$

Case A:	$K_{V\theta}$ =	$0.2\omega_S$
Case B:	$K_{V\theta}$ =	$0.2\omega_S$
Case C:	$K_{V\theta}$ =	$0.4\omega_S$

$$K_{V\theta} \downarrow$$

7a. Determine Rated Flow at System Pressure (Q_{RP})

$$P_S \rightarrow \qquad Q_{RP} = Q_R \sqrt{\frac{P_S}{1000}}$$

$$Q_{RP} \downarrow \qquad\qquad K_{V\theta} \downarrow$$

8a. Determine Position Error due to Valve Uncertainties ($\Delta\theta_U$)

$$D_M \rightarrow \qquad \Delta\theta_U = 0.04 \left[\frac{Q_{RP}}{K_{V\theta}\frac{D_M}{2\pi}} \right]$$

$$Q_{RP} \downarrow \qquad\qquad \Delta\theta_U \downarrow \qquad\qquad K_{V\theta} \downarrow$$

A. POSITION CONTROL SYSTEMS:

9a. Determine Position Error due to External Torque ($\Delta\theta_E$)

Q_{RP} $\Delta\theta_U$ $K_{V\theta}$

$D_M \rightarrow$
$P_S \rightarrow$
$T_L \rightarrow$

$$\Delta\theta_E \;=\; 0.02\left[\frac{Q_{RP}}{K_{V\theta}\frac{D_M}{2\pi}}\right]\left[\frac{T_L}{P_S\frac{D_M}{2\pi}}\right]$$

Q_{RP} $\Delta\theta_U$ $\Delta\theta_E$

10a. Determine Total Position Error ($\Delta\theta_{TOT}$)

$\Delta\theta_{FB} \rightarrow$

$$\Delta\theta_{TOT} \;=\; \Delta\theta_U + \Delta\theta_E + \Delta\theta_{FB}$$

B. VELOCITY CONTROL SYSTEMS:

6b. Determine Open Loop Gain of System ($K_{V\omega}$)

ω_S

Case A: $K_{V\omega} \;=\; 0.2\omega_S$

Case B: $K_{V\omega} \;=\; 0.2\omega_S$

Case C: $K_{V\omega} \;=\; 0.4\omega_S$

$K_{V\omega}$

B. VELOCITY CONTROL SYSTEMS:

7b. Determine Following Error (ω_F)

$$K_{V\omega}$$

$\alpha \rightarrow$

$$\omega_F = \frac{\alpha}{K_{V\omega}}$$

C. PRESSURE CONTROL SYSTEMS:

6c. Determine Open Loop Gain of System (K_{VP})

$$\omega_S$$

Case A: $K_{VP} = 0.2\omega_S$

Case B: $K_{VP} = 0.2\omega_S$

Case C: $K_{VP} = 0.4\omega_S$

$$K_{VP}$$

7c. Determine Pressure Error due to Valve Uncertainties (ΔP_U)

$C_{TOT} \rightarrow$
$D_M \rightarrow$
$Q_{RP} \rightarrow$

$$\Delta P_U = 0.04 \left[\frac{C_{TOT} Q_{RP}}{K_{VP} \frac{D_M}{2\pi}} \right]$$

$\Delta P_U \downarrow$ $K_{VP} \downarrow$

C. PRESSURE CONTROL SYSTEMS:

8c. Determine Pressure Error due to Valve Leakage (ΔP_L)

$$\Delta P_L = 0.02 \left[\frac{C_{TOT} Q_{RP}}{K_{VP}\left(\frac{D_M}{2\pi}\right)^2} \right]\left(\frac{\Delta P_{AB}}{P_S}\right)$$

Inputs: ΔP_U, K_{VP}, C_{TOT}, D_M, Q_{RP}, ΔP_{AB}, P_S

Outputs: ΔP_U, ΔP_L, K_{VP}

9c. Determine Pressure Error due to Motion of Actuator (ΔP_v)

$$\Delta P_V = V\left[\frac{C_{TOT}}{K_{VP}\frac{D_M}{2\pi}} \right]$$

Inputs: C_{TOT}, D_M, V

Outputs: ΔP_V, ΔP_U, ΔP_L, K_{VP}

10c. Determine Total Pressure Error (ΔP_{TOT})

$$\Delta P_{TOT} = \Delta P_U + \Delta P_L + \Delta P_V + \Delta P_{FB}$$

Input: ΔP_{FB}

Output: K_{VP}

C. PRESSURE CONTROL SYSTEMS:

11c. Determine Pressure Following Error (P_F)

$$K_{VP}$$
$$\downarrow$$

$$\Delta P/\Delta t \longrightarrow$$

$$P_F = \frac{\frac{\Delta P}{\Delta t}}{K_{VP}}$$

Symbology

SYMBOL	DESCRIPTION	UNITS
a	Linear Acceleration	in/sec^2
A_1	Piston Area (full bore side)	in^2
A_2	Piston Area (rod end side)	in^2
C_H	Hydraulic Stiffness (Linear)	lb_f/in
C_H	Hydraulic Stiffness (Rotary)	$lb_f\,in/radian$
C_{TOT}	Total Actuator Stiffness (Linear)	lb_f/in
C_{TOT}	Total Actuator Stiffness (Rotary)	$lb_f\,in/radian$
X_0	Cylinder Stroke for Minimum Stiffness	in
D_M	Hydraulic Motor Displacement	in^3/rev
E	Fluid Bulk Modulus	lb_f/in^2
f_V	Servo Valve Natural Frequency	Hz
F_E	External Disturbing Force	lb_f
G_{AMP}	Amplifier Gain (Proportional)	ma/V
G_{AMP}	Amplifier Gain (Integrating)	(ma/sec)/V
G_{SV}	Servo Valve Flow Gain	$(in^3/sec)/ma$
H_{FB}	Feedback Gain (Linear Position)	V/in
H_{FB}	Feedback Gain (Rotary Position)	V/radian
H_{FB}	Feedback Gain (Linear Velocity)	V/(in/sec)
H_{FB}	Feedback Gain (Rotary Velocity)	V/(rad/sec)
H_{FB}	Feedback Gain (Pressure)	V/psi
I_A	Actual Electrical Input Current	% of max.
J_{EFF}	Effective Rotary Load Inertia	$in\,lb_f\,sec^2$
K_V	Open Loop Gain (Velocity Constant)	sec^{-1}
K_{VP}	Open Loop Gain (Pressure)	sec^{-1}

SYMBOL	DESCRIPTION	UNITS
K_{VV}	Open Loop Gain (Linear Velocity)	sec^{-1}
K_{VX}	Open Loop Gain (Linear Position)	sec^{-1}
$K_{V\theta}$	Open Loop Gain (Rotary Position)	sec^{-1}
$K_{V\omega}$	Open Loop Gain (Rotary Velocity)	sec^{-1}
M	Effective Load Mass	$lb_f\,sec^2/in$
N	Motor Speed	rev/min
P_F	Following Error (Pressure)	psi
P_S	System Supply Pressure	psi
$\Delta P/\Delta t$	Rate of Change of Pressure	psi/sec
ΔP_{AB}	Pressure Drop Between Ports A & B	psi
ΔP_L	Pressure Error Due to Leakage	psi
ΔP_v	Pressure Error Due to Load Velocity	psi
ΔP_{TOT}	Total Pressure Error	psi
ΔP_U	Pressure Error From Valve Uncertainty	psi
ΔP_{FB}	Pressure Error Due to Feedback Device	psi
Q_R	Rated Flow at Rated Pressure (1000 psi)	in^3/sec
Q_{RP}	Rated Flow at Operating Pressure	in^3/sec
Q	Flow	GPM
R	Cylinder Area Ratio (A_1/A_2)	(unitless)
S	Maximum Cylinder Stroke	in
T_E	External Disturbing Torque	$lb_f\,in$
V	Linear Velocity	in/sec
V_F	Following Error (Velocity)	in/sec

SYMBOL	DESCRIPTION	UNITS
V_{L1}	Line Volume – Valve to Actuator (A_1)	in^3
V_{L2}	Line Volume – Valve to Actuator (A_2)	in^3
$\Delta V / \Delta t$	Rate of Change of Velocity (= a)	(in/sec)/sec
W_L	Load Weight (Mass x gravity)	lb_f
X_F	Following Error (Position)	in
$\Delta X / \Delta t$	Rate of Change of Position (= V)	in/sec
ΔX_E	Position Error Due to External Force	in
ΔX_{FB}	Position Error Due to Feedback Device	in
ΔX_{TOT}	Total Position Error	in
ΔX_U	Position Error From Valve Uncertainty	in
α	Rotary Acceleration	rad/sec^2
ξ	Damping Coefficient	(unitless)
θ_F	Following Error (Rotary Position)	radians
$\Delta \theta_E$	Rotary Error Due to External Torque	radians
$\Delta \theta_{TOT}$	Total Rotary Error	radians
$\Delta \theta_U$	Rotary Error Due to Valve Uncertainty	radians
ω	Rotary Velocity	rad/sec
ω_F	Following Error (Rotary Velocity)	rad/sec
ω_L	Natural Frequency of Actuator & Load	rad/sec
ω_S	Natural Frequency of System	rad/sec
ω_V	Natural Frequency of Valve	rad/sec

Derivation of Pressure Drop Equations

Derivation of Equations for P_1 and P_2 in Chapter 4, page 97

1. $P_S - P_1 = (P_2 - P_T)R^2$

$$\frac{P_S - P_1}{R^2} = P_2 - P_T$$

$$P_2 = P_T + \frac{P_S - P_1}{R^2}$$

2. $P_1 A_1 = P_2 A_2 + F$
   - Substitute Equation 1 for P_2:

$$P_1 A_1 = \left[P_T + \frac{P_S - P_1}{R^2}\right]A_2 + F$$

   - Divide both sides by A_1:

$$P_1 = \left[P_T + \frac{P_S - P_1}{R^2}\right]\frac{A_2}{A_1} + \frac{F}{A_1}$$

   - $R = \dfrac{A_1}{A_2}$, so $\dfrac{1}{R} = \dfrac{A_2}{A_1}$:

$$P_1 = \left[P_T + \frac{P_S - P_1}{R^2}\right]\frac{1}{R} + \frac{F}{A_1}$$

$$P_1 = \frac{P_T}{R} + \frac{P_S}{R^3} - \frac{P_1}{R^3} + \frac{F}{A_1}$$

   - Make terms Factorable by $\dfrac{1}{R^3}$:

$$P_1 = \frac{P_T R^2}{R^3} + \frac{P_S}{R^3} - \frac{P_1}{R^3} + F\frac{R^3}{A_1 R^3}$$

   - Factor by $\dfrac{1}{R^3}$:

$$P_1 = \frac{1}{R^3}\left[P_T R^2 + P_S - P_1 + \frac{F R^3}{A_1}\right]$$

   - Substitute $R = \dfrac{A_1}{A_2}$:

$$P_1 = \frac{1}{R^3}\left[P_T R^2 + P_S - P_1 + \frac{F R^2 A_1}{A_1 A_2}\right]$$

$$P_1 = \frac{1}{R^3}\left[P_T R^2 + P_S - P_1 + \frac{F R^2}{A_2}\right]$$

- Consolidate P_1 terms on left side:

$$P_1\left[1 + \frac{1}{R^3}\right] = \frac{1}{R^3}\left[P_TR^2 + P_S + \frac{FR^2}{A_2}\right]$$

- Isolate P_1 term on left side:

$$P_1 = \frac{1}{\left[1 + \frac{1}{R^3}\right]}\frac{1}{R^3}\left[P_TR^2 + P_S + \frac{FR^2}{A_2}\right]$$

- Simplify

$$P_1 = \frac{1}{1 + R^3}\left[P_TR^2 + P_S + \frac{FR^2}{A_2}\right]$$

$$P_1 = \frac{1}{1 + R^3}\frac{A_2}{A_2}\left[P_TR^2 + P_S + \frac{FR^2}{A_2}\right]$$

$$P_1 = \frac{1}{A_2(1 + R^3)}\left[P_SA_2 + P_TR^2A_2 + FR^2\right]$$

$$P_1 = \frac{P_SA_2 + R^2(P_TA_2 + F)}{A_2(1 + R^3)}$$

$$P_1 = \frac{P_SA_2 + R^2(F + P_TA_2)}{A_2(1 + R^3)}$$

Servo Catalog Data

SM4 series

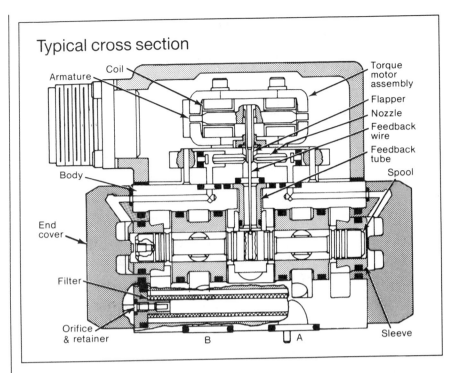

Typical cross section

Coil
Armature
Torque motor assembly
Flapper
Nozzle
Feedback wire
Feedback tube
Body
Spool
End cover
Filter
Orifice & retainer
Sleeve
B
A

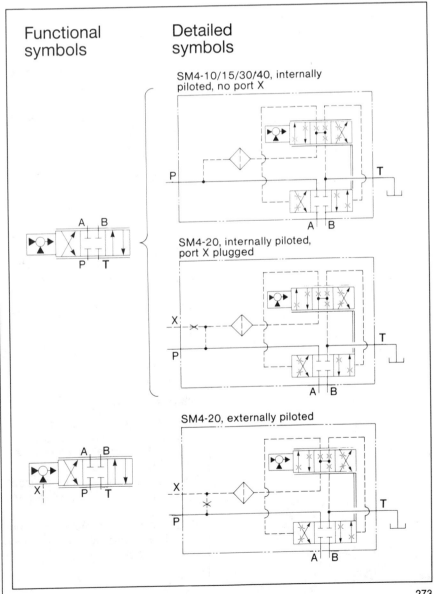

Functional symbols

Detailed symbols

SM4-10/15/30/40, internally piloted, no port X

A B
P T

SM4-20, internally piloted, port X plugged

A B
X P T

SM4-20, externally piloted

A B
X P T

Understanding servo valves for closed loop control

Functional servo diagram

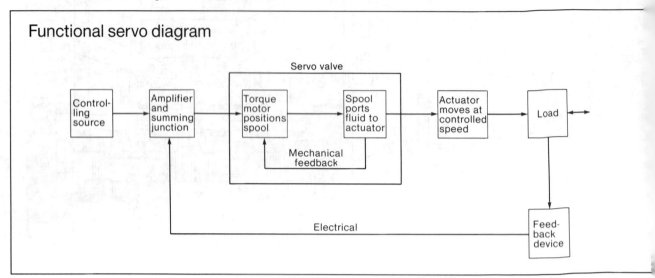

A servo valve is a closed center, four-way proportional flow control.

An electrical input to the first-stage torque motor positions the second-stage spool to control flow proportional to input current.

The second-stage spool position is achieved by internal force feedback.

Excellent hysteresis and linearity results in repeatable and accurate actuator velocity to a commanded position.

System control is completed through an electrical feedback device to the summing junction and servo amplifier.

The Vickers SM4 servo valve and how it works

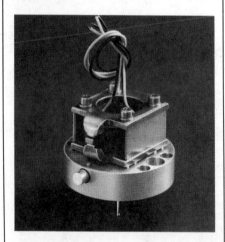

The two stage servo valve with force feedback is capable of performance superior to that of any other design.

Meticulous assembly and measured quality ensure fast, dependable electrohydraulic amplification.

SM4 first stage

The torque motor converts a low level electrical current signal to rotary motion of the armature and flapper assembly.

The pilot stage directs pressurized hydraulic flow as a result of flapper motion.

Pole pieces and permanent magnets provide a network for the magnetic flux in each air gap.

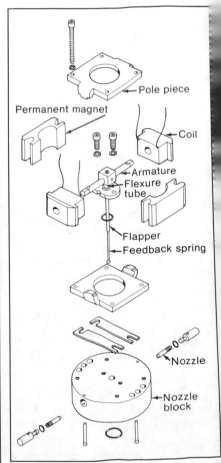

Current applied to the coils creates a torque on the armature.

The nozzle block allows for precision spacing of the control nozzles to the flapper.

The servo valve first stage is an electrical controller with a hydraulic amplifier and force feedback.

First stage operation

The first stage controller converts low level current signals to a mechanical force or motion.

The flexure tube supports the armature and also acts as a fluid seal between hydraulic and electrical sections.

Attached at the center of the armature are the flapper and feedback spring that extend down through the flexure tube.

An input signal is applied to the coils through an electrical connector, polarizing the armature ends and creating rotational torque on the armature.

The flexure tube acts as a spring, limiting the flapper motion between two nozzle openings.

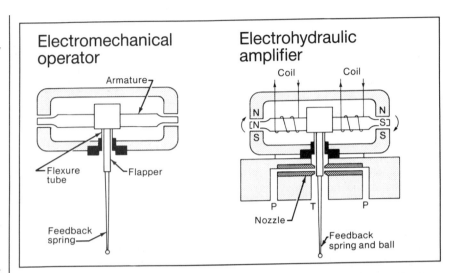

Hydraulic amplification results as pilot flow from P is supplied through an integral filter and orifice assembly to the nozzles for control of a greater second stage flow proportional to input current.

Internal feedback is achieved by the use of a simple cantilever spring attached to the flapper, with the ball end closely fitted to the second stage spool.

The servo valve second stage controls hydraulic flow and contains the pilot stage filter.

M4 second stage

Output flow is controlled by a four-way closed center spool that slides within a sleeve.

Spool movement uncovers openings in the sleeve to meter flow to the control ports.

The positioning of the spool relative to these metering slots provides precise control flow.

The modular pilot stage is precision mounted to the body for positioning the feedback ball in the spool.

The null adjust pin and self locking nut allow for dependable sleeve adjustment of control flow around the center or null position.

A fine mesh filter is located inside the second stage body to protect the pilot stage from contamination.

Various spool/sleeve sizes allow for the proper selection of rated flow and performance.

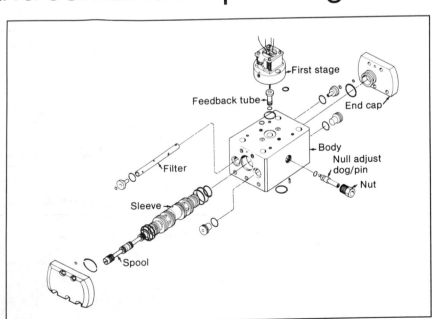

Spool position and output flow are proportional to the level of input current.

Actual load flow to the actuator depends on supply pressure, load pressure drop, input current, and the flow rating or size of the servo valve.

Complete servo valve control flow for optimum system performance

Second stage operation

Electrical current causes the flapper to move toward the nozzle on the right side.

Pilot pressure on the spool end area causes spool movement to the left and control flow out of port A.

The feedback spring bends and applies a force to the flapper, which tends to recenter the flapper between the nozzles.

Spool positioning occurs at the point at which the spring feedback force equals the torque motor force induced by the input current.

The spool stops at this position, and the flapper is now centered until the input current changes to a new level.

A reverse electrical input current signal results in flow to port B.

Optimum flow control is achieved by force feedback.

With constant supply pressure and flow to the servo valve, output control flow is infinitely proportional to the input current.

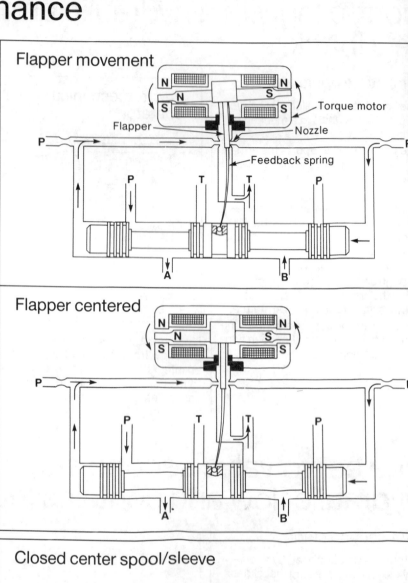

Flapper movement

Flapper centered

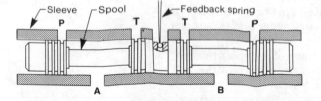

Closed center spool/sleeve

Spool centered at null with the control lands blocking A and B

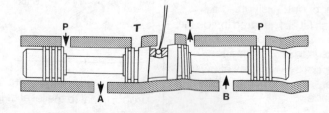

Movement of spool

Spool motion opens flow passage from P to A and B to R

SM4 operating data

All data is typical, based on a large sample of actual tests at 30 cST (141 SUS) and 48°C (120°F).

Rated flow, and total internal leakage at null

Standard model	Rated flow*		Total null leakage	
	l/min	USgpm	l/min	USgpm
SM4-10	38	10	0,95	0.25
SM4-15	57	15	0,95	0.25
SM4-20	76	20	0,95	0.25
SM4-30	113	30	1,50	0.40
SM4-40	151	40	1,50	0.40

Flow gain, null region	50% to 200% nominal within ±5% rated current

Flow gain, normal region for standard models

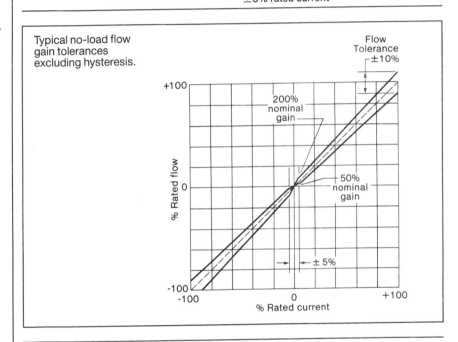

Typical no-load flow gain tolerances excluding hysteresis.

Hysteresis around null	< 2% of rated current
Symmetry error	< 5% of rated current
Linearity error	< 5% of rated current and <10% at maximum rated flow
Threshold	<0.5% of rated current

*Maximum flow at 70 bar (1000 psi) Δp and typical leakage at 210 bar (3000 psi).

Change in rated flow with valve pressure drop

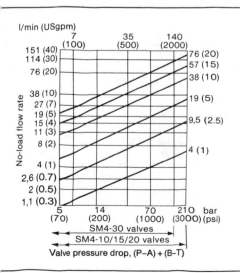

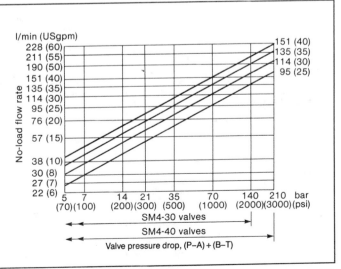

Rated supply pressure:	
SM4-10/15/20/40	210 bar (3000 psi)
SM4-30	140 bar (2000 psi)
Maximum supply pressure*:	
SM4-10/15/20/40	350 bar (5000 psi)
SM4-30	210 bar (3000 psi)
*Subject to engineering approval	
Minimum supply pressure	14 bar (200 psi)
Proof pressure:	
At supply port	150% max. rated pressure
At return port	100% max. rated pressure
Burst pressure, return port open	250% max. rated pressure

Pressure gain –
the change of load pressure drop with input current, with no valve flow and closed control ports.

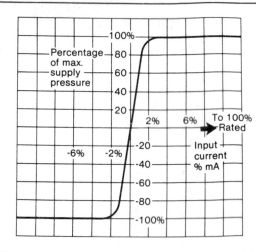

Pressure gain, null region:	>30% of supply pressure per 1% of rated current
Null shift, nominal:	
– temperature change	<1.5%/38°C (100°F)
– supply pressure change	< 2%/70 bar (1000 psi)
– return pressure change	< 2%/35 bar (500 psi)
– acceleration	< 2%/10g
Vibration tested, 5 Hz to 2000 Hz at 10g and double amplitude along each axis	No damage to components
Shock tested, up to 150g along all axes	No damage to components
Endurance tested, to ISO 6404	No degradation from standard valve performance limits
Hydraulic fluids, temperature ranges, and filtration recommendations	See page 55
Installation (start-up)	See page 55
Installation dimensions	See pages 24 and 25
Mass (weight):	
SM4-10/15	0,68 kg (1.5 lb)
SM4-20	1,05 kg (2.3 lb)
SM4-30	1,9 kg (4.1 lb)
SM4-40	2,8 kg (6.2 lb)

Power transmission efficiency – the maximum power envelope expressed as a percentage.

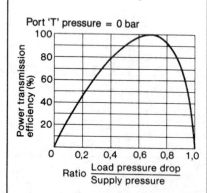

Servo performance is optimum when valve pressure drop is one third of supply pressure. Consider overall hydraulic efficiency in sizing system heat exchangers.

SM4 performance data

All data is typical with actual performance limits defined by Vickers test procedure TP7841 or TP7795.

Frequency response

Is the relationship of no-load control flow to input current with a sinusoidal current sweep at constant amplitude over a range of frequencies. Expressed in frequency (Hz), amplitude ratio (dB), and phase shift (degrees).

3db ratio is the standard comparison point, and 90 degree phase shift is a measure of the servo valve bandwidth.

Frequency response is lower for increased valve size because of changes in internal design, i. e. spool/sleeve diameters, flapper nozzle assembly, feedback spring rates.

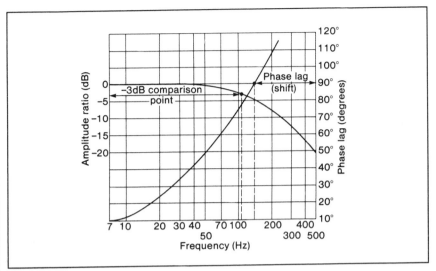

Vickers SM4 torque motors are magnetically stabilized for reliable servo valve performance from 14 – 210 bar (200 – 3000 psi) operating pressures.

Typical frequency response curves for standard models

Note: SM4-10/15/20/40 shown at 210 bar (3000 psi), SM4-30 at 140 bar (2000 psi)

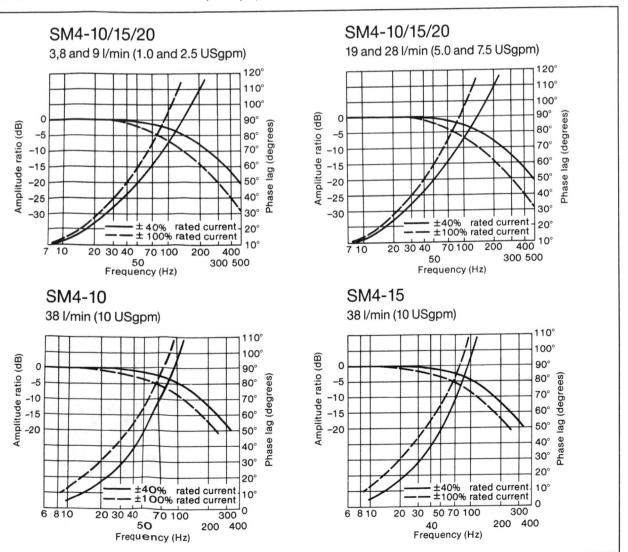

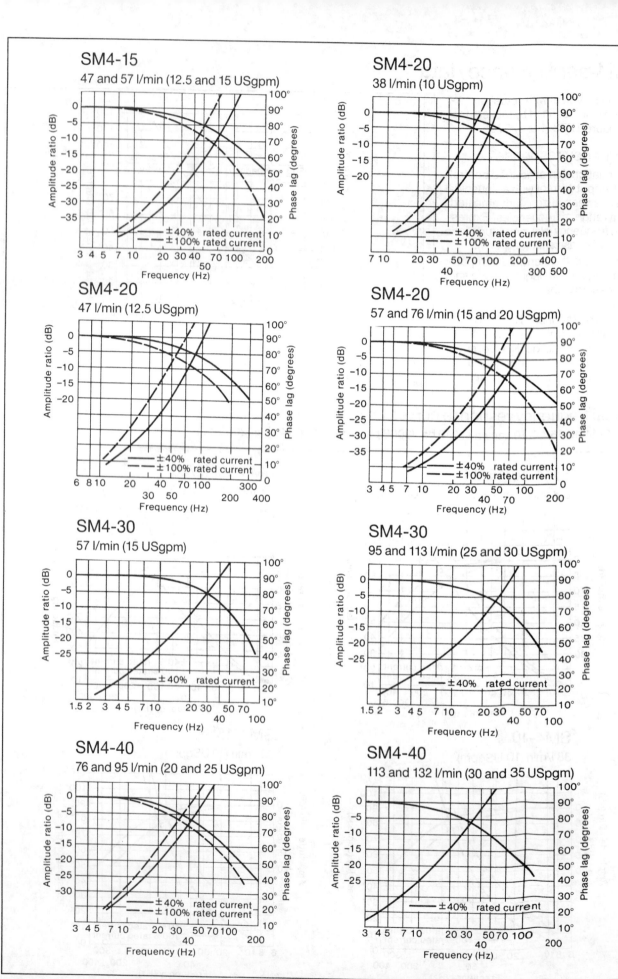

SM4 performance data

Typical frequency response curves

SM4-40
151 l/min (40 USgpm)

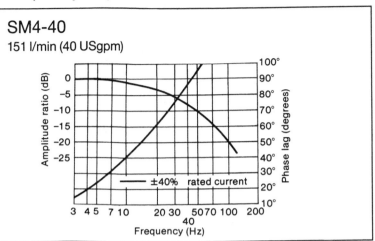

Changes in frequency response with pressure

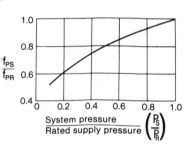

f_{PS} = frequency @ 90° phase lag at any pressure P
f_{PR} = frequency @ 90° phase lag at rated pressure P_R

SM4-10/15/20/40 P_R = 210 bar (3000 psi)
SM4-30 P_R = 140 bar (2000 psi)

Example:
Assuming a 151 l/min (40 USgpm) SM4-40 valve is to be used at 165 bar (2400 psi) pressure P, variation in response with supply pressure =

$$\left(\frac{P_S}{P_R}\right) = 0.8; \text{ then } \frac{f_{PS}}{f_{PR}} = 0.92.$$

From curve above, f_{PR} = 45 Hz; therefore, f_{PS} = 45(0.92) = 40 Hz.

Step response

Step response – the typical rise time to achieve a percentage of control flow output. Settling time is that in which transient flow fluctuations diminish to a stated level.

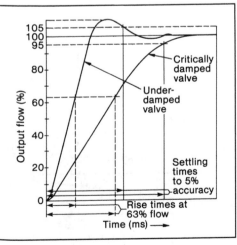

Typical step response for standard models

Note: SM4-10/15/20/40 shown at 210 bar (3000 psi), SM4-30 at 140 bar (2000 psi)

Typical step response curves

SM4-10/15/20
3,8 and 9 l/min (1.0 and 2.5 USgpm)

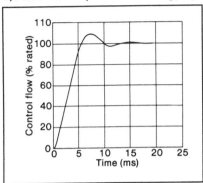

SM4-10/15/20
19 and 28 l/min (5.0 and 7.5 USgpm)

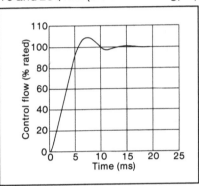

SM4-10/15/20
38 l/min (10 USgpm)

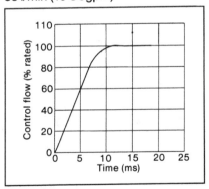

SM4-15/20
47 and 57 l/min (12.5 and 15 USgpm)

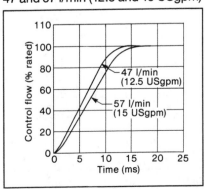

Typical step response curves

SM4-20
76 l/min (20 USgpm)

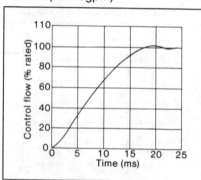

SM4-30
57, 95, and 113 l/min
(15, 25 and 30 USgpm)

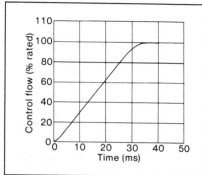

SM4-40
76, 95, and 113 l/min
(20, 25, and 30 USgpm)

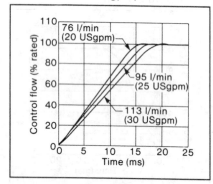

SM4-40
132 and 151 l/min (35 and 40 USgpm)

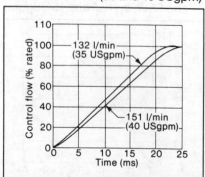

SM4 electrical data

Select coil resistance and connections for compatible interface to servo electronics.

Recommended coil resistance is shown in bold print.

Standard Coil resistance selection		
Nominal resistance per coil at 21°C (70°F)	Rated current mA	
	Single, parallel or differential	Series connection
Ohms	mA	mA
20	**200**	**100**
80	**40**	**20**
30	100	50
200	15	7.5
Optional		
80	50	25
140	40	20
200	20	10
300	30	15
1000	10	5
1500	8	4

Example of recommended coil configurations:

Characteristic	Single	Series	Parallel	Differenti
Coil resistance (Ohms)	80	160	40	80
Rated current (mA)	± 40	± 20	± 40	± 40
Nominal voltage (Volts)	± 3.2	± 3.2	± 1.6	± 3.2
Approx. inductance (Henrys)	0.22	0.66	0.18	0.3

The two coils of the valve can be connected in any of the ways shown in the diagram below varying the assembly inside the mating connector (Amphenol No. MS3016A-14S-2S).

Electrical polarity for control flow out of port B

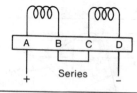

Series:
A+, D−
Connect B and C.

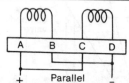

Parallel:
A+, C+
B−, D−
Connect A and C.
Connect B and D.

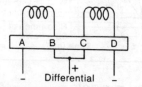

Differential:
A−, D−
B+, C+
Connect B and C.
BC−, current BA > CD
BC+, current CD > BA

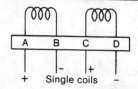

Single:
A+, B−
or
C+, D−

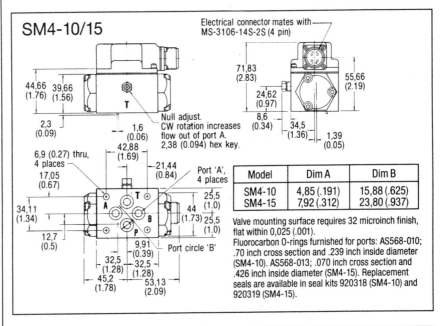

SM4 installation dimensions; mm (inches)

SM4-10/15

Electrical connector mates with MS-3106-14S-2S (4 pin)

Null adjust. CW rotation increases flow out of port A. 2,38 (0.094) hex key.

44,66 (1.76) 39,66 (1.56)

2,3 (0.09)   1,6 (0.06)   42,88 (1.69)

6,9 (0.27) thru, 4 places

17,05 (0.67)   21,44 (0.84)   Port 'A', 4 places

34,11 (1.34)   25,5 (1.0)   44 (1.73)   25,5 (1.0)

12,7 (0.5)   9,91 (0.39)   Port circle 'B'

32,5 (1.28)   32,5 (1.28)

45,2 (1.78)   53,13 (2.09)

71,83 (2.83)   55,66 (2.19)

24,62 (0.97)   8,6 (0.34)   34,5 (1.36)   1,39 (0.05)

Model	Dim A	Dim B
SM4-10	4,85 (.191)	15,88 (.625)
SM4-15	7,92 (.312)	23,80 (.937)

Valve mounting surface requires 32 microinch finish, flat within 0,025 (.001).

Fluorocarbon O-rings furnished for ports: AS568-010; .70 inch cross section and .239 inch inside diameter (SM4-10). AS568-013; .070 inch cross section and .426 inch inside diameter (SM4-15). Replacement seals are available in seal kits 920318 (SM4-10) and 920319 (SM4-15).

SM4-20

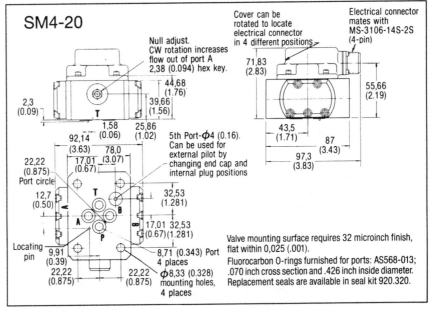

Null adjust. CW rotation increases flow out of port A. 2,38 (0.094) hex key.

Cover can be rotated to locate electrical connector in 4 different positions

Electrical connector mates with MS-3106-14S-2S (4-pin)

44,68 (1.76)   39,66 (1.56)

2,3 (0.09)   1,58 (0.06)   25,86 (1.02)

92,14 (3.63)   78,0 (3.07)

22,22 (0.875)   17,01 (0.67)   Port circle

12,7 (0.50)   32,53 (1.281)

17,01 32,53 (0.67)(1.281)

Locating pin   9,91 (0.39)   8,71 (0.343) Port 4 places

22,22 (0.875)   22,22 (0.875)   ⌀8,33 (0.328) mounting holes, 4 places

5th Port-⌀4 (0.16). Can be used for external pilot by changing end cap and internal plug positions

71,83 (2.83)   55,66 (2.19)

43,5 (1.71)   87 (3.43)

97,3 (3.83)

Valve mounting surface requires 32 microinch finish, flat within 0,025 (.001).

Fluorocarbon O-rings furnished for ports: AS568-013; .070 inch cross section and .426 inch inside diameter. Replacement seals are available in seal kit 920.320.

Activation of optional 5th port (SM4-20) ▶ Procedure

1. Remove end caps (24) (will be relocated to opposite ends of valve in step 5).
2. Remove plug (36) and o-rings (35), (37).
3. Remove orifice (32), filter screen (34) and o-rings (31), (32). Install these items in place of items removed in step 2.
4. Install items removed in step 2 where orifice and filter screen were previously located. Reference step 3.
5. Install valve and caps on the opposite ends of valve from which they were removed. Reference step 1.
6. The valve's 5th port is now activated

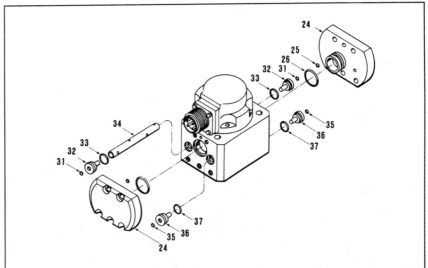

SM4-30

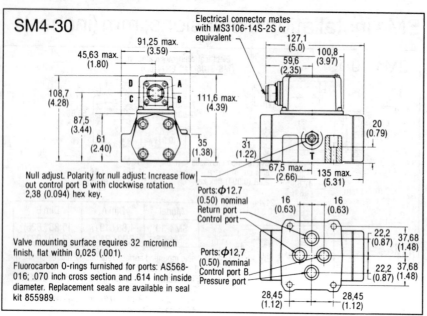

Electrical connector mates with MS3106-14S-2S or equivalent

91,25 max. (3.59)

45,63 max. (1.80)

108,7 (4.28)

D A
C B

111,6 max. (4.39)

87,5 (3.44)

61 (2.40)

35 (1.38)

127,1 (5.0)

59,6 (2.35)

100,8 (3.97)

20 (0.79)

31 (1.22)

T

67,5 max. (2.66)

135 max. (5.31)

Null adjust. Polarity for null adjust: Increase flow out control port B with clockwise rotation. 2,38 (0.094) hex key.

Valve mounting surface requires 32 microinch finish, flat within 0,025 (.001).

Fluorocarbon O-rings furnished for ports: AS568-016; .070 inch cross section and .614 inch inside diameter. Replacement seals are available in seal kit 855989.

Ports: Φ12.7 (0.50) nominal
Return port
Control port

Ports: Φ12,7 (0.50) nominal
Control port B
Pressure port

16 (0.63) 16 (0.63)

22,2 (0.87) 37,68 (1.48)

22,2 (0.87) 37,68 (1.48)

28,45 (1.12) 28,45 (1.12)

SM4-40

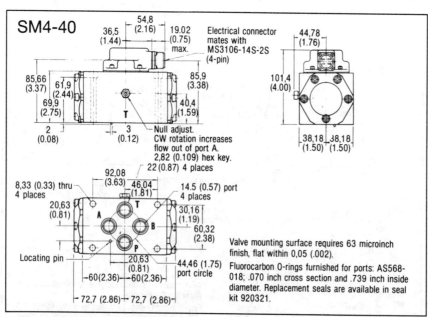

54,8 (2.16)

36,5 (1.44)

19.02 (0.75) max.

Electrical connector mates with MS3106-14S-2S (4-pin)

44,78 (1.76)

85,66 (3.37)

61,9 (2.44)

69,9 (2.75)

85,9 (3.38)

40,4 (1.59)

T

2 (0.08) 3 (0.12)

Null adjust. CW rotation increases flow out of port A. 2,82 (0.109) hex key. 22 (0.87) 4 places

101,4 (4.00)

38,18 (1.50) 38,18 (1.50)

92,08 (3.63)

46,04 (1.81)

8,33 (0.33) thru 4 places

20,63 (0.81)

A T
B
P

14.5 (0.57) port 4 places

30,16 (1.19)

60,32 (2.38)

Locating pin

20,63 (0.81)

60 (2.36) 60 (2.36)

72,7 (2.86) 72,7 (2.86)

44,46 (1.75) port circle

Valve mounting surface requires 63 microinch finish, flat within 0,05 (.002).

Fluorocarbon O-rings furnished for ports: AS568-018; .070 inch cross section and .739 inch inside diameter. Replacement seals are available in seal kit 920321.

SM4 selection guidelines

Sizing and application examples to simplify the selection of servo valves

Vickers servo calculations are valuable in initial sizing of system components, but full static and dynamic analyses should be made for each application. Please consult your Vickers representative.

Principles of servo system analysis:

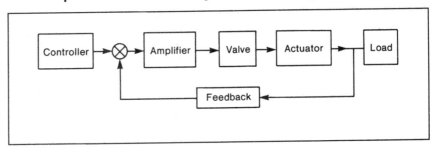

A servo system is a set of components that includes a servo valve, analog command source, power amplifier/summing junction, actuator, and feedback transducer.

In a closed loop servo system, the output of the actuator or load is continuously measured by a feedback transducer and compared to the command input signal.

Any errors are amplified and the resultant signal is applied to the servo valve to correct the actuator and load disturbance.

These devices have static and dynamic characteristics which can influence the selection of the proper servo valve.

In order to calculate the servo valve rated flow, it is important to consider the physical configuration of the actuator and load, total force requirements, maximum valocity, and acceleration/deceleration limits. Total force, F, includes all forces due to acceleration/deceleration, friction, and other external forces.

The following 12 application examples show typical configurations for differential (unequal area) cylinders, symmetrical (equal area) cylinders, and hydraulic motors. The force is considered for positive (resistive) and negative (overrunning) loads.

Type of feedback

Position	$\pm$ cm ($\pm$ in)
Velocity	$\pm$ cm/s ($\pm$ in/sec)
	or $\pm$ r/min ($\pm$ rev/min)
Pressure	$\pm$ bar (psi)
Force	$\pm$ daN (lbf)

Application examples of hydraulic cylinders and motors with positive and negative load configurations

1.0 Extending differential cylinder with a positive load.

1.1 Retracting differential cylinder with a positive load.

2.0 Extending differential cylinder with a negative load.

2.1 Retracting differential cylinder with a negative load.

3.0 Extending differential cylinder on an inclined plane with a positive load.

3.1 Retracting differential cylinder on an inclined plane with a positive load.

4.0 Extending differential cylinder on an inclined plane with a negative load.

4.1 Retracting differential cylinder on an inclined plane with a negative load.

5.0 Extend/retract symmetrical cylinder with a positive load.

5.1 Extend/retract symmetrical cylinder with a negative load.

6.0 Rotate a hydraulic motor with a positive load.

6.1 Rotate a hydraulic motor with a negative load.

Application notes:

1. If extend and retract load velocity requirements are the same, then select the servo valve rated flow based on the retract example.

2. If the calculated load pressure is more than the supply pressure, the load may not retract.

3. If the calculated load pressure is a negative value, then cavitation may occur.

4. If necessary, change the cylinder or motor size, supply pressure, and/ or velocity requirements.

5. For all examples, calculate the servo valve rated flow based on the given parameters and then optimize the system variables by recalculation as desired.

Given (or calculated) parameters for cylinder application examples

The ideal electrohydraulic cylinder provides hydraulic force proportional to servo valve differential pressure, and velocity proportional to servo valve control flow.

Parameter or nomenclature	Calculation or symbol	Metric units	English units	Conversion
Acceleration (or deceleration)	$a = \dfrac{v_{MAX}}{t_a}$	$\dfrac{cm}{s^2}$	$\dfrac{in}{sec^2}$	$2.54 \dfrac{cm}{s^2} \Big/ \dfrac{in}{sec^2}$
Area, cap end	$A_1 = \dfrac{\pi D_P^2}{4}$	cm^2	in^2	
Area, rod end	$A_2 = \dfrac{\pi}{4}(D_P^2 - D_R^2)$	cm^2	in^2	$6.45\ cm^2/in^2$
Area ratio	$R = \dfrac{A_1}{A_2}$			
Diameter, piston	D_P	cm	in	2.54 cm/in
Diameter, rod	D_R	cm	in	
Force, total required	$F = F_a + F_E + F_C + F_S$	daN	lbf	0.445 daN/lbf
Force, acceleration	$F_a = Ma$			
Force, additive due to external disturbances	F_E			
Force, load friction $F_C = \mu\,W_L$ for horizontal plane $F_C = \mu\,W_L \cos\theta$ for inclined plane where μ = coefficient of friction with typical values of 0.1 to 0.3 θ = angle of incline in degrees				
Force, maximum cylinder	$F_{MAX} = A_1 \cdot P_S$			
Force, seal friction	$F_S = 0.1\,F_{MAX}$ or use values from the applicable cylinder catalog			
Load flow	$Q_L = 0,06(A)v_{MAX}$ when $A = cm^2$ $v_{MAX} = cm/sec$ $Q_L = \dfrac{A \cdot v_{MAX}}{3.85}$ when $A = in^2$ $v_{MAX} = in/sec$	l/min	USgpm	$\dfrac{3.78\ l/min}{USgpm}$
Load weight	W_L	daN	lbf	0.445 daN/lb
Mass, total	$M = \dfrac{W_L}{g} + M_P$ where $g = 1000$ cm/sec^2 or $g = 386$ in/sec^2	kg $\left(daN\dfrac{sec^2}{cm}\right)$	lb $\left(lbf\dfrac{sec^2}{in}\right)$	0.454 kg/lb
Mass, piston	M_P = values from applicable cylinder catalog	kg	lb	

Parameter or nomenclature	Calculation or symbol	Metric units	English units	Conversion
Pressure, cap end	P_1 = calculated for each example			
rod end	P_2 = calculated for each example	bar	psi	0.070 bar/psi
supply	P_S = based on cylinder or power unit size			
return (tank)	P_T = dependent on return line restrictions			
Stroke, cylinder	S	cm	in	2.54 cm/in
Time, acceleration deceleration total limit	t_a t_d t_l	sec	sec	
Velocity, maximum cylinder	$v_{MAX} = \dfrac{S}{(t_l - t_a)}$ where $t_a = t_d$	$\dfrac{cm}{sec}$	$\dfrac{in}{sec}$	$2.54\,\dfrac{cm}{sec}\Big/\dfrac{in}{sec}$

Given (or calculated) parameters for hydraulic motor application examples

The ideal electrohydraulic motor provides torque proportional to servo valve differential pressure, and velocity (speed) proportional to servovalve control flow.

Parameter or nomenclature	Calculation or symbol	Metric units	English units	Conversion
Acceleration, angular (or deceleration)	$a = \dfrac{\omega_M}{t_a}$	$\dfrac{radians}{sec^2}$	$\dfrac{radians}{sec^2}$	
Displacement (motor)	D_M	$\dfrac{cm^3}{r}$	$\dfrac{in^3}{rev}$	$\dfrac{16.4\ cm^3/r}{in^3/rev}$
Inertia, total applied. Inertia, load Inertia, motor	Jeff $= J_L + J_M$ J_L $J_M = .005\,J_L$ to $.02\,J_L$ or use values from the applicable motor catalog	Nm-sec²	lbfin-sec²	$\dfrac{0.113\ Nm\text{-}sec^2}{lbfin\text{-}sec^2}$
Load flow (motor)	$Q_{ML} = \dfrac{n_M \cdot D_M}{1000}$ when n_M = rev/min D_M = cm³/rev $Q_{ML} = \dfrac{n_M \cdot D_M}{231}$ when n_M = rev/min D_M = in³/rev	l/min USgpm		$\dfrac{3.78\ l/min}{USgpm}$
Pressure, inlet to motor	P_1 = calculated for each example			
Pressure, outlet from motor	P_2 = calculated for each example	bar	psi	$\dfrac{0.070\ bar}{psi}$
Pressure, supply	P_S = based on load torque required			
Pressure, tank (return)	P_T = dependent on motor P required			
Rotation, motor Speed (velocity), maximum motor	θ_M $\omega_M = \dfrac{\theta_M}{t_l}$	radians $\dfrac{radians}{sec}$	radians $\dfrac{radians}{sec}$	
Motor speed	$n_M = \dfrac{60\,\omega_M}{2\pi}$	$\dfrac{rev}{min}$	$\dfrac{rev}{min}$	$9{,}55\,\dfrac{rad}{sec}\Big/\dfrac{rev}{min}$
Time, acceleration deceleration total limit	t_a t_d t_l	sec sec sec	sec sec sec	
Torque, total required	$T = \alpha\,Jeff + T_L = T_D$	Nm	lbf·in	$\dfrac{0.113\ Nm}{lbf\cdot in}$
Torque, damping	T_D = use values from applicable motor catalog or estimate $T_D = 0.10\,T_L$ to $0.15\,T_L$			
Torque, load	T_L			

Application example 1.0

Differential cylinder extending with a positive load.

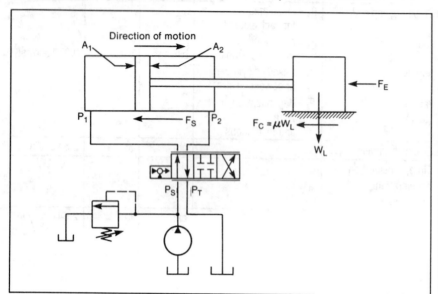

Configuration 1.0:

$$F = Ma + F_C + F_E + F_S \qquad \text{daN (lbf)}$$

Using given parameters, find P_1 and P_2.

$$P_1 = \frac{P_S A_2 + R^2(F + P_T A_2)}{A_2(1 + R^3)} \qquad \text{bar (psi)}$$

$$P_2 = P_T + \frac{P_S - P_1}{R^2} \qquad \text{bar (psi)}$$

Check cylinder sizing and calculate rated flow, Q_R, dependent on load pressure P_1.

$$Q_L = 0.06\,(A_1)\,v_{MAX} \qquad \text{l/min}$$

$$Q_L = \frac{(A_1)\,v_{MAX}}{3.85} \qquad \text{USgpm}$$

$$Q_R = Q_L \sqrt{\frac{35}{P_S - P_1}} \qquad \text{l/min}$$

$$Q_R = Q_L \sqrt{\frac{500}{P_S - P_1}} \qquad \text{USgpm}$$

Select a standard SM4 servo valve size equal to or greater than the calculated Q_R.

Given parameters 1.0:
in Metric units

$$
\begin{aligned}
F &= 4450 \text{ daN} \\
P_S &= 210 \text{ bar} \\
P_T &= 5{,}25 \text{ bar} \\
A_1 &= 53{,}5 \text{ cm}^2 \\
A_2 &= 38{,}1 \text{ cm}^2 \\
R &= 1{,}4 \\
v_{MAX} &= 30 \text{ cm/s}
\end{aligned}
$$

Calculations 1.0:
in Metric units

$$P_1 = \frac{210\,(38{,}1) + 1{,}4^2\,(4450 + 5{,}25\,[38{,}1])}{38{,}1\,(1 + 1{,}4^3)} = 120$$

$$P_1 = 120 \text{ bar}$$

$$P_2 = 5{,}25 + \frac{210 - 120}{1{,}4^2} = 50$$

$$P_2 = 50 \text{ bar}$$

$$Q_L = 0{,}06\,(53{,}5)\,30 = 96$$

$$Q_L = 96 \text{ l/min}$$

$$Q_R = 96 \sqrt{\frac{35}{210 - 120}} = 60$$

$$Q_R = 60 \text{ l/min}$$

Given parameters 1.0:
in English units

$$
\begin{aligned}
F &= 10{,}000 \text{ lbf} \\
P_S &= 3{,}000 \text{ psi} \\
P_T &= 75 \text{ psi} \\
A_1 &= 8.3 \text{ in}^2 \\
A_2 &= 5.9 \text{ in}^2 \\
R &= 1.4 \\
v_{MAX} &= 12 \text{ in/sec}
\end{aligned}
$$

Calculations 1.0:
in English units

$$P_1 = \frac{3000\,(5.9) + 1.4^2\,(10{,}000 + 75\,[5.9])}{5.9\,(1 + 1.4^3)} = 1727$$

$$P_1 = 1727 \text{ psi}$$

$$P_2 = 75 + \frac{3000 - 1727}{1.4^2} = 724$$

$$P_2 = 724 \text{ psi}$$

$$Q_L = \frac{(8.3)\,12}{3.85} = 26$$

$$Q_L = 26 \text{ USgpm}$$

$$Q_R = 26 \sqrt{\frac{500}{3000 - 1727}} = 16$$

$$Q_R = 16 \text{ USgpm}$$

Application example 1.1
Differential cylinder retracting with a positive load.

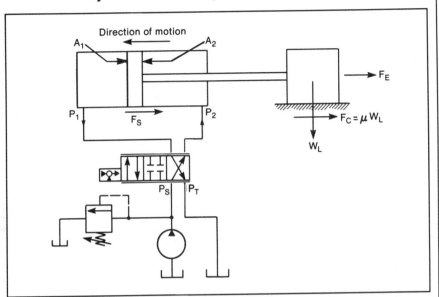

Given parameters 1.1:
in English units

F	=	10,000 lbf
P_S	=	3,000 psi
P_T	=	75 psi
A_1	=	8.3 in²
A_2	=	5.9 in²
R	=	1.4
v_{MAX}	=	12 in/sec

Calculations 1.1:
in English units

$$P_2 = \frac{3000\,(5.9)\,1.4^3 + 10,000 + 75\,(5.9)\,1.4}{5.9\,(1 + 1.4^3)} = 2678$$

$P_2 = 2678$ psi

$P_1 = 75 + (3000 - 2678)\,1.4^2 = 706$

$P_1 = 706$ psi

$$Q_L = \frac{(5.9)\,12}{3.85} = 18$$

$Q_L = 18$ USgpm

$$Q_R = 18\sqrt{\frac{500}{3000 - 2678}} = 22$$

$Q_R = 22$ USgpm

Configuration 1.1:

$F = Ma + F_C + F_E + F_S \qquad$ daN (lbf)

Using given parameters,
find P_2 and P_1.

$$P_2 = \frac{P_S A_2 R^3 + (F + P_T A_2 R)}{A_2(1 + R^3)} \quad \text{bar (psi)}$$

$P_1 = P_T + (P_S - P_2)\,R^2 \qquad$ bar (psi)

Check cylinder sizing and calculate rated flow, Q_R, dependent on load pressure P_2.

$Q_L = 0{,}06\,(A_2)\,v_{MAX} \qquad$ l/min

$Q_L = \dfrac{(A_2)\,v_{MAX}}{3.85} \qquad$ USgpm

$Q_R = Q_L\sqrt{\dfrac{35}{P_S - P_2}} \qquad$ l/min

$Q_R = Q_L\sqrt{\dfrac{500}{P_S - P_2}} \qquad$ USgpm

Select a standard SM4 servo valve size equal to or greater than the calculated Q_R.

Given parameters 1.1:
in Metric units

F	=	4450 daN
P_S	=	210 bar
P_T	=	5,25 bar
A_1	=	53,5 cm²
A_2	=	38,1 cm²
R	=	1,4
v_{MAX}	=	30 cm/s

Calculations 1.1:
in Metric units

$$P_2 = \frac{210\,(38,1)\,1,4^3 + 4450 + 5,25\,(38,1)\,1,4}{38,1\,(1 + 1,4^3)} = 187$$

$P_2 = 187$ bar

$P_1 = 5{,}25 + (210 - 187)\,1{,}4^2 = 50$

$P_1 = 50$ bar

$Q_L = 0{,}06\,(38{,}1)\,30 = 69$

$Q_L = 69$ l/min

$$Q_R = 69\sqrt{\frac{35}{210 - 187}} = 85$$

$Q_R = 85$ l/min

Application example 2.0

Differential cylinder extending with a negative load.

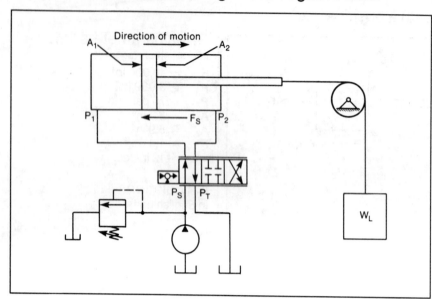

Given parameters 2.0:
in English units

$$F = -5{,}000 \text{ lbf}$$
$$P_S = 2{,}500 \text{ psi}$$
$$P_T = 0 \text{ psi}$$
$$A_1 = 12.6 \text{ in}^2$$
$$A_2 = 9.5 \text{ in}^2$$
$$R = 1.3$$
$$v_{MAX} = 5 \text{ in/sec}$$

Calculations 2.0:
in English units

$$P_1 = \frac{2500\,(9.5) + 1.3^2\,(-5000 + 0\,[9.5])}{9.5\,(1 + 1.3^3)} = 503$$

$$P_1 = 503 \text{ psi}$$

$$P_2 = 0 + \frac{2500 - 503}{1.3^2} = 1181$$

$$P_2 = 1181 \text{ psi}$$

$$Q_L = \frac{(12.6)\,5}{3.85} = 16$$

$$Q_L = 16 \text{ USgpm}$$

$$Q_R = 16 \sqrt{\frac{500}{2500 - 503}} = 8.0$$

$$Q_R = 8.0 \text{ USgpm}$$

Configuration 2.0:

$$F = Ma + F_S - W_L \qquad \text{daN (lbf)}$$

Using given parameters, find P_1 and P_2.

$$P_1 = \frac{P_S A_2 + R^2 (F + P_T A_2)}{A_2 (1 + R^3)} \qquad \text{bar (psi)}$$

$$P_2 = P_T + \frac{P_S - P_1}{R^2} \qquad \text{bar (psi)}$$

Check cylinder sizing and calculate servo valve rated flow, Q_R, dependent on load pressure P_1.

$$Q_L = 0{,}06\,(A_1)\,v_{MAX} \qquad \text{l/min}$$

$$Q_L = \frac{(A_1)\,v_{MAX}}{3.85} \qquad \text{USgpm}$$

$$Q_R = Q_L \sqrt{\frac{35}{P_S - P_1}} \qquad \text{l/min}$$

$$Q_R = Q_L \sqrt{\frac{500}{P_S - P_1}} \qquad \text{USgpm}$$

Select a standard SM4 servo valve size equal to or greater than the calculated Q_R.

Given parameters 2.0:
in Metric units

$$F = -2225 \text{ daN}$$
$$P_S = 175 \text{ bar}$$
$$P_T = 0 \text{ bar}$$
$$A_1 = 81{,}3 \text{ cm}^2$$
$$A_2 = 61{,}3 \text{ cm}^2$$
$$R = 1{,}3$$
$$v_{MAX} = 12{,}7 \text{ cm/s}$$

Calculations 2.0:
in Metric units

$$P_1 = \frac{175\,(61{,}3) + 1{,}3^2\,(-2225 + 0\,[61{,}3])}{61{,}3\,(1 + 1{,}3^3)} = 36$$

$$P_1 = 36 \text{ bar}$$

$$P_2 = 0 + \frac{175 - 36}{1{,}3^2} = 82$$

$$P_2 = 82 \text{ bar}$$

$$Q_L = 0{,}06\,(81{,}3)\,12{,}7 = 62$$

$$Q_L = 62 \text{ l/min}$$

$$Q_R = 62 \sqrt{\frac{35}{175 - 36}} = 31$$

$$Q_R = 31 \text{ l/min}$$

Application example 2.1

Differential cylinder retracting with a negative load.

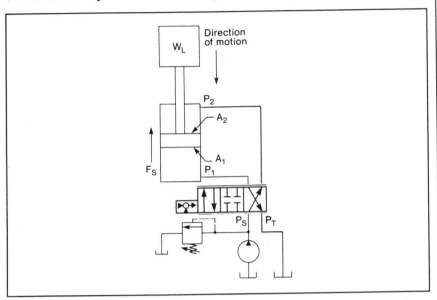

Given parameters 2.1:
in English units

F = –10,000 lbf
P_S = 3,000 psi
P_T = 0 psi
A_1 = 12.6 in^2
A_2 = 9.5 in^2
R = 1.3
v_{MAX} = 10 in/sec

Calculations 2.1:
in English units

$$P_2 = \frac{3000\,(9.5)\,1.3^3 - 10,000 + 0\,(9.5)\,1.3}{9.5\,(1 + 1.3^3)} = 1734$$

$$P_2 = 1734 \text{ psi}$$

$$P_1 = 0 + (3000 - 1734)\,1.3^2 = 2140$$

$$P_1 = 2140 \text{ psi}$$

$$Q_L = \frac{(9.5)\,10}{3.85} = 25$$

$$Q_L = 25 \text{ USgpm}$$

$$Q_R = 25 \sqrt{\frac{500}{3000 - 1734}} = 16$$

$$Q_R = 16 \text{ USgpm}$$

Configuration 2.1:

$$F = Ma + F_S - W_L \qquad \text{daN (lbf)}$$

Using given parameters,
find P_2 and P_1.

$$P_2 = \frac{P_S A_2 R^3 + (F + P_T A_2 R)}{A_2(1 + R^3)} \qquad \text{bar (psi)}$$

$$P_1 = P_T + (P_S - P_2)\,R^2 \qquad \text{bar (psi)}$$

Check cylinder sizing and calculate
servo valve rated flow, Q_R, dependent
on cap end pressure P_2.

$$Q_L = 0,06\,(A_2)\,v_{MAX} \qquad \text{l/min}$$

$$Q_L = \frac{(A_2)\,v_{MAX}}{3.85} \qquad \text{USgpm}$$

$$Q_R = Q_L \sqrt{\frac{35}{P_S - P_2}} \qquad \text{l/min}$$

$$Q_R = Q_L \sqrt{\frac{500}{P_S - P_2}} \qquad \text{USgpm}$$

Select a standard SM4 servo valve
size equal to or greater than the calculated Q_R.

Given parameters 2.1:
in Metric units

F = –4450 daN
P_S = 210 bar
P_T = 0 bar
A_1 = 81,3 cm^2
A_2 = 61,3 cm^2
R = 1,3
v_{MAX} = 25,4 cm/s

Calculations 2.1:
in Metric units

$$P_2 = \frac{210\,(61,3)\,1,3^3 - 4450 + 0\,(61,3)\,1,3}{61,3\,(1 + 1,3^3)} = 122$$

$$P_2 = 122 \text{ bar}$$

$$P_1 = 0 + (210 - 122)\,1,3^2 = 149$$

$$P_1 = 149 \text{ bar}$$

$$Q_L = 0,06\,(61,3)\,25,4 = 93$$

$$Q_L = 93 \text{ l/min}$$

$$Q_R = 93 \sqrt{\frac{35}{210 - 122}} = 59$$

$$Q_R = 59 \text{ l/min}$$

Application example 3.0

Differential cylinder extending on an inclined plane with a positive load.

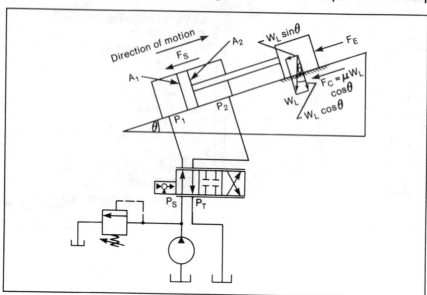

Configuration 3.0:

$$F = Ma + F_E + F_S + W_L (\mu \cos\theta + \sin\theta) \text{ daN (lbf)}$$

Using given parameters,
find P_1 and P_2.

$$P_1 = \frac{P_S A_2 + R^2 (F + P_T A_2)}{A_2 (1 + R^3)} \quad \text{bar (psi)}$$

$$P_2 = P_T + \frac{P_S - P_1}{R^2} \quad \text{bar (psi)}$$

Check cylinder sizing and calculate servo valve rated flow, Q_R, dependent on cap end pressure P_1.

$$Q_L = 0.06 (A_1) v_{MAX} \quad \text{l/min}$$

$$Q_L = \frac{(A_1) v_{MAX}}{3.85} \quad \text{USgpm}$$

$$Q_R = Q_L \sqrt{\frac{35}{P_S - P_1}} \quad \text{l/min}$$

$$Q_R = Q_L \sqrt{\frac{500}{P_S - P_1}} \quad \text{USgpm}$$

Select a standard SM4 servo valve size equal to or greater than the calculated Q_R.

Given parameters 3.0:
in Metric units

F	=	2225 daN
P_S	=	140 bar
P_T	=	3,5 bar
A_1	=	31,6 cm²
A_2	=	19,9 cm²
R	=	1,6
v_{MAX}	=	12,7 cm/s

Calculations 3.0:
in Metric units

$$P_1 = \frac{140 (19,9) + 1,6^2 (2225 + 3,5 [19,9])}{19,9 (1 + 1,6^3)} = 85$$

$$P_1 = 85 \text{ bar}$$

$$P_2 = 3,5 + \frac{140 - 85}{1,6^2} = 25$$

$$P_2 = 25 \text{ bar}$$

$$Q_L = 0,06 (31,6) 12,7 = 24$$

$$Q_L = 24 \text{ l/min}$$

$$Q_R = 24 \sqrt{\frac{35}{140 - 85}} = 19$$

$$Q_R = 19 \text{ l/min}$$

Given parameters 3.0:
in English units

F	=	5,000 lbf
P_S	=	2,000 psi
P_T	=	50 psi
A_1	=	4.9 in²
A_2	=	3.1 in²
R	=	1.6
v_{MAX}	=	5 in/sec

Calculations 3.0:
in English units

$$P_1 = \frac{2000 (3.1) + 1.6^2 (5,000 + 50 [3.1])}{3.1 (1 + 1.6^3)} = 1228$$

$$P_1 = 1228 \text{ psi}$$

$$P_2 = 50 + \frac{2000 - 1228}{1.6^2} = 352$$

$$P_2 = 352 \text{ psi}$$

$$Q_L = \frac{(4.9) 5}{3.85} = 6.4$$

$$Q_L = 6.4 \text{ USgpm}$$

$$Q_R = 6.4 \sqrt{\frac{500}{2000 - 1228}} = 5.2$$

$$Q_R = 5.2 \text{ USgpm}$$

Application example 3.1

Differential cylinder retracting on an inclined plane with a positive load.

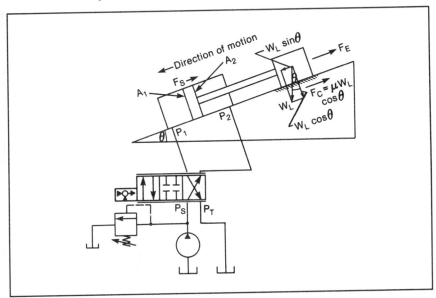

Configuration 3.1:

$$F = Ma + F_E + F_S + W_L(\mu \cos\theta - \sin\theta) \text{ daN (lbf)}$$

Using given parameters, find P_2 and P_1.

$$P_2 = \frac{P_S A_2 R^3 + F + P_T A_2 R}{A_2(1 + R^3)} \quad \text{bar (psi)}$$

$$P_1 = P_T + (P_S - P_2) R^2 \quad \text{bar (psi)}$$

Check cylinder sizing and calculate servo valve rated flow, Q_R, dependent on rod end load pressure P_2.

$$Q_L = 0.06 (A_2) v_{MAX} \quad \text{l/min}$$

$$Q_L = \frac{(A_2) v_{MAX}}{3.85} \quad \text{USgpm}$$

$$Q_R = Q_L \sqrt{\frac{35}{P_S - P_2}} \quad \text{l/min}$$

$$Q_R = Q_L \sqrt{\frac{500}{P_S - P_2}} \quad \text{USgpm}$$

Select a standard SM4 servo valve size equal to or greater than the calculated Q_R.

Given parameters 3.1:
in Metric units

F	=	1780 daN
P_S	=	140 bar
P_T	=	3,5 bar
A_1	=	31,6 cm²
A_2	=	19,9 cm²
R	=	1,6
v_{MAX}	=	12,7 cm/s

Calculations 3.1:
in Metric units

$$P_2 = \frac{140\,(19,9)\,1,6^3 + 1780 + 3,5\,(19,9)\,1,6}{19,9\,(1 + 1,6^3)} = 131$$

$$P_2 = 131 \text{ bar}$$

$$P_1 = 3,5 + (140 - 131)\,1,6^2 = 26$$

$$P_1 = 26 \text{ bar}$$

$$Q_L = 0,06\,(19,9)\,12,7 = 15$$

$$Q_L = 15 \text{ l/min}$$

$$Q_R = 15 \sqrt{\frac{35}{140 - 131}} = 30$$

$$Q_R = 30 \text{ l/min}$$

Given parameters 3.1:
in English units

F	=	4,000 lbf
P_S	=	2,000 psi
P_T	=	50 psi
A_1	=	4.9 in²
A_2	=	3.1 in²
R	=	1.6
v_{MAX}	=	5 in/sec

Calculations 3.1:
in English units

$$P_2 = \frac{2000\,(3.1)\,1.6^3 + 4,000 + 50\,(3.1)\,1.6}{3.1\,(1 + 1.6^3)} = 1874$$

$$P_2 = 1874 \text{ psi}$$

$$P_1 = 50 + (2000 - 1874)\,1.6^2 = 373$$

$$P_1 = 373 \text{ psi}$$

$$Q_L = \frac{(3.1)\,5}{3.85} = 4.0$$

$$Q_L = 4.0 \text{ USgpm}$$

$$Q_R = 4.0 \sqrt{\frac{500}{2000 - 1874}} = 8.0$$

$$Q_R = 8.0 \text{ USgpm}$$

Application example 4.0

Differential cylinder extending on an inclined plane with a negative load.

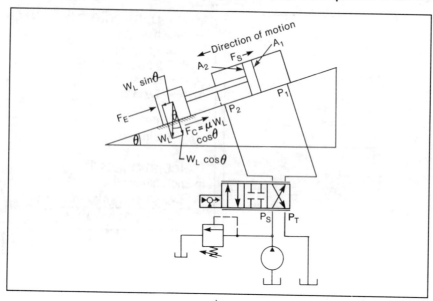

$P_2 = 116$ bar

$Q_L = 0{,}06\ (126)\ 25{,}4 = 192$

$Q_L = 192$ l/min

$Q_R = 192\ \sqrt{\dfrac{35}{210-44}} = 88$

$Q_R = 88$ l/min

Given parameters 4.0:
in English units

F   = −15,000 lbf
P_S = 3,000 psi
P_T = 0 psi
A_1 = 8.3 in^2
A_2 = 5.9 in^2
R   = 1.4
v_{MAX} = 10 in/sec

Calculations 4.0:
in English units

$$P_1 = \frac{3000\,(5.9) + 1.4^2\,(-15{,}000 + 0\,[5.9])}{5.9\,(1 + 1.4^3)} = -530$$

$P_1 = -530$ psi

Caution:

Negative load will cause cylinder cavitation. Change given parameters by increasing cylinder size, system pressure, or decreasing the total force required.

choose; in English units

$A_1 = 19.6$ in^2
$A_2 = 16.5$ in^2
R = 1.2

$$P_1 = \frac{3000\,(16.5) + 1.2^2\,(-15{,}000 + 0\,[16.5])}{16.5\,(1 + 1{,}2^3)} = 620$$

$P_1 = 620$ psi

$$P_2 = 0 + \frac{3000 - 620}{1.2^2} = 1653$$

$P_2 = 1653$ psi

$$Q_L = \frac{(19.6)\ 10}{3.85} = 51$$

$Q_L = 51$ USgpm

$$Q_R = 51\ \sqrt{\frac{500}{3000 - 620}} = 23$$

$Q_R = 23$ USgpm

Configuration 4.0:

$$F = Ma + F_E + F_S + W_L\ (\mu \cos\theta - \sin\theta)\ \text{daN (lbf)}$$

Using given parameters, find P_1 and P_2.

$$P_1 = \frac{P_S A_2 + R^2 (F + P_T A_2)}{A_2 (1 + R^3)} \quad \text{bar (psi)}$$

$$P_2 = P_T + \frac{P_S - P_1}{R^2} \quad \text{bar (psi)}$$

Check cylinder sizing and calculate servo valve rated flow, Q_R, dependent on cap end pressure P_1.

$$Q_L = 0{,}06\ (A_1)\ v_{MAX} \quad \text{l/min}$$

$$Q_L = \frac{(A_1)\ v_{MAX}}{3.85} \quad \text{USgpm}$$

$$Q_R = Q_L\ \sqrt{\frac{35}{P_S - P_1}} \quad \text{l/min}$$

$$Q_R = Q_L\ \sqrt{\frac{500}{P_S - P_1}} \quad \text{USgpm}$$

Select a standard SM4 servo valve size equal to or greater than the calculated Q_R.

Given parameters 4.0:
in Metric units

F   = −6675 daN
P_S = 210 bar
P_T = 0 bar
A_1 = 53,5 cm^2
A_2 = 38,1 cm^2
R   = 1,4
v_{MAX} = 25,4 cm/s

Calculations 4.0:
in Metric units

$$P_1 = \frac{210\,(38{,}1) + 1{,}4^2\,(-6675 + 0\,[38{,}1])}{38{,}1\,(1 + 1{,}4^3)} = -36$$

$P_1 = -36$ bar

Caution:

Negative load will cause cylinder cavitation. Change given parameters by increasing cylinder size, system pressure, or decreasing the total force required.

choose; in Metric units

$A_1 = 126$ cm^2
$A_2 = 106$ cm^2
R = 1,2

Calculations 4.0:
in Metric units

$$P_1 = \frac{210\,(106) + 1{,}2^2\,(-6675 + 0\,[106])}{106\,(1 + 1{,}2^3)} = 44$$

$P_1 = 44$ bar

$$P_2 = \frac{210 - 44}{1{,}2^2} = 116$$

Application example 4.1
Differential cylinder retracting on an inclined plane with a negative load.

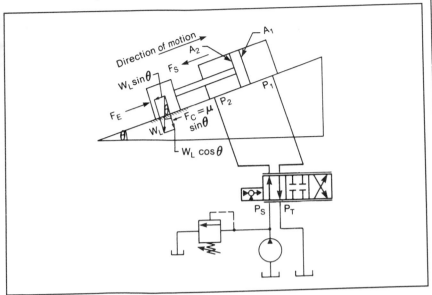

Given parameters 4.1:
in English units

F	=	−15,000 lbf
P_S	=	3,000 psi
P_T	=	0 psi
A_1	=	8.3 in^2
A_2	=	5.9 in^2
R	=	1.4
v_{MAX}	=	10 in/sec

Calculations 4.1:
in English units

$$P_2 = \frac{3000\,(5.9)\,1.4^3 - 15,000 + 0\,(5.9)\,1.4}{5.9\,(1 + 1.4^3)} = 1518$$

$P_2 = 1518$ psi

$P_1 = 0 + (3000 - 1518)\,1.4^2 = 2904$

$P_1 = 2904$ psi

$Q_L = \dfrac{(5.9)\,10}{3.85} = 15$

$Q_L = 15$ USgpm

$Q_R = 15\,\sqrt{\dfrac{500}{3000 - 1518}} = 8.7$

$Q_R = 8.7$ USgpm

Configuration 4.1:

$F = Ma + F_E + F_S + W_L\,(\mu\cos\theta + \sin\theta)$ daN (lbf)

Using given parameters,
find P_2 and P_1.

$P_2 = \dfrac{P_S A_2 R^3 + F + P_T A_2 R}{A_2(1 + R^3)}$ bar (psi)

$P_1 = P_T + (P_S - P_2)\,R^2$ bar (psi)

Check cylinder sizing and calculate
servo valve rated flow, Q_R, dependent
on rod end pressure P_2.

$Q_L = 0,06\,(A_2)\,v_{MAX}$ l/min

$Q_L = \dfrac{(A_2)\,v_{MAX}}{3.85}$ USgpm

$Q_R = Q_L\,\sqrt{\dfrac{35}{P_S - P_2}}$ l/min

$Q_R = Q_L\,\sqrt{\dfrac{500}{P_S - P_2}}$ USgpm

Select a standard SM4 servo valve
size equal to or greater than the calculated Q_R.

Given parameters 4.1:
in Metric units

F	=	−6675 daN
P_S	=	210 bar
P_T	=	0 bar
A_1	=	53,5 cm^2
A_2	=	38,1 cm^2
R	=	1,4
v_{MAX}	=	25,4 cm/s

Calculations 4.1:
in Metric units

$$P_2 = \frac{210\,(38,1)\,1,4^3 - 6375 + 0\,(38,1)\,1,4}{38,1\,(1 + 1,4^3)} = 107$$

$P_2 = 107$ bar

$P_1 = 0 + (210 - 107)\,1,4^2 = 202$

$P_1 = 202$ bar

$Q_L = 0,06\,(38,1)\,25,4 = 58$

$Q_L = 58$ l/min

$Q_R = 58\,\sqrt{\dfrac{35}{210 - 107}} = 34$

$Q_R = 34$ l/min

Application example 5.0

Symmetrical cylinder with a positive load.

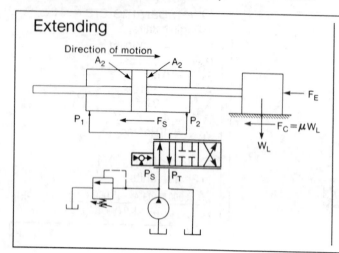

Extending	Retracting

Configuration 5.0:

$$F = Ma + F_C + F_E + F_S \qquad \text{daN (lbf)}$$

Using given parameters, find P_1.

$$P_1 = \frac{P_S A_2 + F + P_T A_2}{2 A_2} \qquad \text{bar (psi)}$$

$$P_2 = P_S - P_1 + P_T \qquad \text{bar (psi)}$$

Check cylinder sizing and calculate servo valve rated flow, Q_R, dependent on pressure P_1.

$$Q_L = 0{,}06 \, (A_2) \, v_{MAX} \qquad \text{l/min}$$

$$Q_L = \frac{(A_2) \, v_{MAX}}{3.85} \qquad \text{USgpm}$$

$$Q_R = Q_L \sqrt{\frac{35}{P_S - P_1}} \qquad \text{l/min}$$

$$Q_R = Q_L \sqrt{\frac{500}{P_S - P_1}} \qquad \text{USgpm}$$

Select a standard SM4 servo valve size equal to or greater than the calculated Q_R.

Given parameters 5.0:
in Metric units

$$
\begin{aligned}
F &= 4450 \text{ daN} \\
P_S &= 175 \text{ bar} \\
P_T &= 5{,}25 \text{ bar} \\
A_2 &= 38{,}1 \text{ cm}^2 \\
v_{MAX} &= 12{,}7 \text{ cm/s}
\end{aligned}
$$

Calculations 5.0:
in Metric units

$$P_1 = \frac{175 \, (38{,}1) + 4450 + 5{,}25 \, (38{,}1)}{2 \, (38{,}1)} = 149$$

$$P_1 = 149 \text{ bar}$$

$$P_2 = 175 - 149 + 5{,}25 = 31$$

$$P_2 = 31 \text{ bar}$$

$$Q_L = 0{,}06 \, (38{,}1) \, 12{,}7 = 29$$

$$Q_L = 29 \text{ l/min}$$

$$Q_R = 29 \sqrt{\frac{35}{175 - 149}} = 34$$

$$Q_R = 34 \text{ l/min}$$

Given parameters 5.0:
in English units

$$
\begin{aligned}
F &= 10{,}000 \text{ lbf} \\
P_S &= 2{,}500 \text{ psi} \\
P_T &= 75 \text{ psi} \\
A_2 &= 5{.}9 \text{ in}^2 \\
v_{MAX} &= 5 \text{ in/sec}
\end{aligned}
$$

Calculations 5.0:
in English units

$$P_1 = \frac{2500 \, (5.9) + 10{,}000 + 75 \, (5.9)}{2 \, (5.9)} = 2134$$

$$P_1 = 2134 \text{ psi}$$

$$P_2 = 2500 - 2134 + 75 = 441$$

$$P_2 = 441 \text{ psi}$$

$$Q_L = \frac{(5.9) \, 5}{3.85} = 7.6$$

$$Q_L = 7.6 \text{ USgpm}$$

$$Q_R = 7.6 \sqrt{\frac{500}{2500 - 2134}} = 9.0$$

$$Q_R = 9.0 \text{ USgpm}$$

Application example 5.1

Symmetrical cylinder with a negative load.

Extending	Retracting
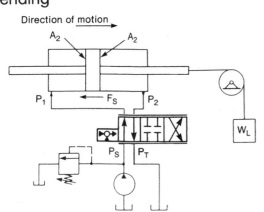	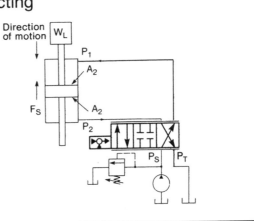

Configuration 5.1:

$$F = Ma + F_S - W_L \quad \text{daN (lbf)}$$

Using given parameters, find P_1.

$$P_1 = \frac{P_S A_2 + F + P_T A_2}{2\,A_2} \quad \text{bar (psi)}$$

$$P_2 = P_S - P_1 + P_T \quad \text{bar (psi)}$$

Check cylinder sizing and calculate servo valve rated flow, Q_R, dependent on pressure P_1.

$$Q_L = 0{,}06\,(A_2)\,v_{MAX} \quad \text{l/min}$$

$$Q_L = \frac{(A_2)\,v_{MAX}}{3.85} \quad \text{USgpm}$$

$$Q_R = Q_L \sqrt{\frac{35}{P_S - P_1}} \quad \text{l/min}$$

$$Q_R = Q_L \sqrt{\frac{500}{P_S - P_1}} \quad \text{USgpm}$$

Select a standard SM4 servo valve size equal to or greater than the calculated Q_R.

Given parameters 5.1:
in Metric units

F	=	-4450 daN
P_S	=	210 bar
P_T	=	0 bar
A_2	=	60,6 cm^2
v_{MAX}	=	12,7 cm/s

Calculations 5.1:
in Metric units

$$P_1 = \frac{210\,(60{,}6) - 4450 + 0\,(60{,}6)}{2\,(60{,}6)} = 68$$

$$P_1 = 68 \text{ bar}$$

$$P_2 = 210 - 68 + 0 = 142$$

$$P_2 = 142 \text{ bar}$$

$$Q_L = 0{,}06\,(60{,}6)\,12{,}7 = 46$$

$$Q_L = 46 \text{ l/min}$$

$$Q_R = 46 \sqrt{\frac{35}{210 - 68}} = 23$$

$$Q_R = 23 \text{ l/min}$$

Given parameters 5.1:
in English units

F	=	$-10,000$ lbf
P_S	=	3,000 psi
P_T	=	0 psi
A_2	=	9.4 in^2
v_{MAX}	=	5 in/sec

Calculations 5.1:
in English units

$$P_1 = \frac{3000\,(9.4) - 10,000 + 0\,(9.4)}{2\,(9.4)} = 968$$

$$P_1 = 968 \text{ psi}$$

$$P_2 = 3000 - 968 + 0 = 2032$$

$$P_1 = 2032 \text{ psi}$$

$$Q_L = \frac{(9.4)\,5}{3.85} = 12$$

$$Q_L = 12 \text{ USgpm}$$

$$Q_R = 12 \sqrt{\frac{500}{3000 - 968}} = 6.0$$

$$Q_R = 6.0 \text{ USgpm}$$

Application example 6.0

Hydraulic motor with a positive load.

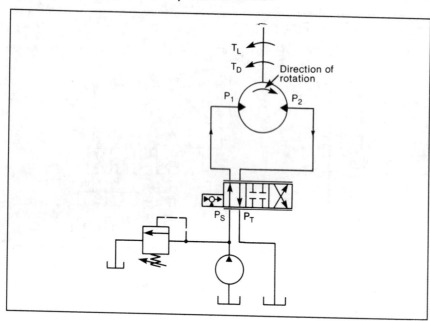

Given parameters 6.0:
in English units

T = 500 lbfin
P_S = 3000 psi
P_T = 0 psi
D_M = 5 in^3/rev.
n_M = 100 rev/min

Calculations 6.0:
in English units

$$P_1 = \frac{3000 + 0}{2} + \frac{\pi\, 500}{5} = 1814$$

P_1 = 1814 psi

P_2 = 3000 − 1814 + 0 = 1186

P_2 = 1186 psi

$$Q_{ML} = \frac{100(5)}{231} = 2.0$$

Q_{ML} = 2.0 USgpm

$$Q_R = 2.0 \sqrt{\frac{500}{3000 - 1814}} = 1.3$$

Q_R = 1.3 USgpm

Configuration 6.0:

$$T = \alpha J_{EFF} + T_D + T_L \quad \text{Nm (lbfin)}$$

Using given parameters, find P_1 and P_2.

$$P_1 = \frac{P_S + P_T}{2} + \frac{10\,\pi T}{D_M} \quad \text{bar}$$

$$P_1 = \frac{P_S + P_T}{2} + \frac{\pi T}{D_M} \quad \text{psi}$$

$$P_2 = P_S - P_1 + P_T \quad \text{bar (psi)}$$

Check cylinder sizing and calculate servo valve rated flow, Q_R, dependent on pressure P_1.

$$Q_{ML} = \frac{n_M \cdot D_M}{1000} \quad \text{l/min}$$

$$Q_{ML} = \frac{n_M \cdot D_M}{231} \quad \text{USgpm}$$

$$Q_R = Q_{ML} \sqrt{\frac{35}{P_S - P_1}} \quad \text{l/min}$$

$$Q_R = Q_{ML} \sqrt{\frac{500}{P_S - P_1}} \quad \text{USgpm}$$

Select a standard SM4 servo valve size equal to or greater than the calculated Q_R.

Given parameters 6.0:
in Metric units

T = 56,5 Nm
P_S = 210 bar
P_T = 0 bar
D_M = 82 cm^3/r
n_M = 100 rev/min

Calculations 6.0:
in Metric units

$$P_1 = \frac{210 + 0}{2} + \frac{10\pi\,(56,5)}{82} = 127$$

P_1 = 127 bar

P_2 = 210 − 127 + 0 = 83

P_2 = 83 bar

$$Q_{ML} = \frac{100(82)}{1000} = 8,2$$

Q_{ML} = 8,2 l/min

$$Q_R = 8,2 \sqrt{\frac{35}{210 - 127}} = 5,3$$

Q_R = 5,3 l/min

Application example 6.1

Hydraulic motor with a negative load.

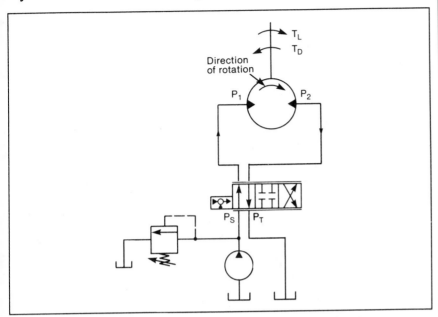

T_L
T_D

Direction of rotation

P_1 P_2

P_S P_T

Given parameters 6.1:
in English units

T = −1500 lbfin
P_S = 3000 psi
P_T = 0 psi
D_M = 5 in³/rev.
n_M = 100 rev/min

Calculations 6.1:
in English units

$$P_1 = \frac{3000 + 0}{2} + \frac{\pi(-1500)}{5} = 558$$

$P_1 = 558$ psi

$P_2 = 3000 - 558 + 0 = 2442$

$P_2 = 2442$ psi

$$Q_{ML} = \frac{100(5)}{231} = 2.0$$

$Q_{ML} = 2.0$ USgpm

$$Q_R = 2.0 \sqrt{\frac{500}{3000 - 558}} = 1.0$$

$Q_R = 1.0$ USgpm

Configuration 6.1:

$T = \alpha J_{EFF} + T_D - T_L$ Nm (lbfin)

Using given parameters,
find P_1 and P_2.

$P_1 = \dfrac{P_S + P_T}{2} + \dfrac{10\,\pi T}{D_M}$ bar

$P_1 = \dfrac{P_S + P_T}{2} + \dfrac{\pi T}{D_M}$ psi

$P_2 = P_S - P_1 + P_T$ bar (psi)

Check cylinder sizing and calculate
servo valve rated flow, Q_R, dependent
on pressure P_1.

$Q_{ML} = \dfrac{n_M \cdot D_M}{1000}$ l/min

$Q_{ML} = \dfrac{n_M \cdot D_M}{231}$ USgpm

$Q_R = Q_{ML} \sqrt{\dfrac{35}{P_S - P_1}}$ l/min

$Q_R = Q_{ML} \sqrt{\dfrac{500}{P_S - P_1}}$ USgpm

Select a standard SM4 servo valve
size equal to or greater than the calculated Q_R.

Given parameters 6.1:
in Metric units

T = −170 Nm
P_S = 210 bar
P_T = 0 bar
D_M = 82 cm³/r
n_M = 100 rev/min

Calculations 6.1:
in Metric units

$$P_1 = \frac{210 + 0}{2} + \frac{10\pi(-170)}{82} = 40$$

$P_1 = 40$ bar

$P_2 = 210 - 40 + 0 = 170$

$P_2 = 170$ bar

$$Q_{ML} = \frac{100(82)}{1000} = 8{,}2$$

$Q_{ML} = 8{,}2$ l/min

$$Q_R = 8{,}2 \sqrt{\frac{35}{210 - 40}} = 3{,}6$$

$Q_R = 3{,}6$ l/min

Ordering information

Specify the following requirements to ensure the proper model selection.

Using the data on catalog pages 277 to 284 and the servo valve application and sizing guidelines on pages 285 to 299, determine whether a standard or customized valve is needed for your application. After selecting the servo valve model and the part number, choose the required accessory products using the charts following each valve size on pages 301 to 303.

Model code for standard SM4 servo valves

SM4-**-(***)***-****/***-**

   ①   ②   ③   ④

① Valve size and port circle mounting interface

Valve size	Port circle
10	15,9 mm (0.625 in)
15	23,9 mm (0.937 in)
20	22,2 mm (0.875 in)
30	* Non-circular
40	44,5 mm (1.750 in)

*See page 284 for port configuration

② Rated flow
at 70 bar (1000 psi) Δp (P-A-B-T)

Code (USgpm) l/min	Available with SM4-				
	10	15	20	30	40
(1.0) 3,8	●	●	●		
(2.5) 9	●	●	●		
(5.0) 19	●	●	●		
(7.5) 28	●	●	●		
(10) 38	●	●	●		
(12.5) 47		●	●		
(15) 57		●	●		
(20) 76			●	●	●
(25) 95				●	●
(30) 113				●	●
(35) 132					●
(40) 151					●

③ Coil resistance/rated current code (Ohms/mA) at 21°C (70°F)

200/15 80/40 30/100 20/100	Standard options for SM4-10, 15, 20 and 40 sizes
80/65	For SM4-30 only

④ Design number

10 series for SM4-10, 15, 20 and 40.
20 series for SM4-30.
Subject to change.
Installation dimension unaltered for design numbers *0 to *9 inclusive

Servo valves and accessories must be ordered separately.

Example of an order for inch units

Quantity	Model description	Part no.
1	SM4-20(10)38-200/15-10 valve	689783
1	Valve mounting bolt kit BK866687	866687
1	SM4M-20-10 rear port subplate manifold	682997
1	Subplate manifold bolt kit BK855992	855992
1	SM4FV-20-10 flushing valve	682999
1	Flushing valve bolt kit	688701

Example of an order for metric units

Quantity	Model description	Part no.
1	SM4-20(10)38-200/15-10 valve	689783
1	Valve mounting bolt kit BK866690M	866690M
1	SM4M-20-10M rear port subplate manifold	866664
1	Subplate manifold bolt kit BK855993M	855993M
1	SM4FV-20-10 flushing valve	682999
1	Flushing valve bolt kit BK689630M	68963M

Model code for customized versions of SM4 servo valves

a. From the standard code above, determine the designation of the servo valve nearest to the requirements.

b. Separately list and define the characteristics of all customized features for the application. Your Vickers representative will be pleased to assist.

c. After completion of any necessary design and development work, Vickers will assign an S-number suffix that will define the group of special features. The complete model designation will then be in the form of;

SM4-**-(***)***-****/***-**-S***

                  ⑤

⑤ Special features suffix

One unique suffix, e. g. S963, will denote a particular group of customized features such as:

Rated flow,
coil resistance/current,
spool % overlap/underlap,
electrical connector position,
5th port (pilot pressure) SM4-20 only,
high frequency response,
high pressure version.

Note: Customized models will affect price and delivery.

Standard model and part numbers are listed on pages 301 to 303. Recommended selections are shown in bold print for your convenience.

Standard SM4-10 servo valves

Model number	Part number
SM4-10(1)3.8-80/40-10	**855356**
SM4-10(1)3.8-20/200-10	**687982**
SM4-10(1)3.8-200/15-10	855352
SM4-10(1)3.8-30/100-10	855357
SM4-10(2.5)9-80/40-10	**689896**
SM4-10(2.5)9-20/200-10	**683582**
SM4-10(2.5)9-200/15-10	855353
SM4-10(2.5)9-30/100-10	855358
SM4-10(5)19-80/40-10	**689875**
SM4-10(5)19-20/200-10	**683005**
SM4-10(5)19-200/15-10	855354
SM4-10(5)19-30/100-10	855359
SM4-10(7.5)28-80/40-10	**514604**
SM4-10(7.5)28-20/200-10	**683565**
SM4-10(7.5)28-200/15-10	855355
SM4-10(7.5)28-30/100-10	855360
SM4-10(10)38-80/40-10	**855970**
SM4-10(10)38-20/200-10	**688727**
SM4-10(10)38-200/15-10	855969
SM4-10(10)38-30/100-10	855971

Recommended selections shown in bold print.

SM4-10 accessories

Description	Model number	Part number
Valve mounting bolt kit 1/4 – 20 x 2 1/4"	BK866685	866685
Valve mounting bolt kit M6 x 60 mm	BK689623M	689623
Subplate, rear port (inch)	SM4M-10-10	682142
Subplate, rear port (metric)	SM4M-10-10M	866654
Subplate, side port (inch)	SM4ME-10-10	682143
Subplate, side port (metric)	SM4ME-10-10M	866656
Subplate mounting bolt kit 1/4 – 20 x 1 1/2"	BK855992	855992
Subplate mounting bolt kit M6 x 40 mm	BK855993M	855993
Adaptor to CETOP-3 (inch)	SM4A-3-10-10	686617
Adaptor to CETOP-3 (metric)	SM4A-3-10-10M	866655
Adaptor-3 mounting bolt kit 10 – 24 x 1/2"	BK855984	855984
Adaptor-3 mounting bolt kit M5 x 12 mm	BK855985M	855985
Flushing valve (inch)	SM4FV-10/15-10	633231
Flushing valve (metric)	SM4FV-10/15-10M	633231
Flushing valve mounting bolt kit 1/4–20 x 1"	BK866686	866686
Flushing valve mounting bolt kit M6 x 25 mm	BK689629M	689629
Cable connector	MS3106-14S-2S	242123
Cable clamp	MS3057-6	126058

Standard SM4-15 servo valves

Model number	Part number
SM4-15(1)3.8-80/40-10	**855366**
SM4-15(1)3.8-20/200-10	**855374**
SM4-15(1)3.8-200/15-10	855361
SM4-15(1)3.8-30/100-10	855358
SM4-15(2.5)9-80/40-10	**635664**
SM4-15(2.5)9-20/200-10	**683017**
SM4-15(2.5)9-200/15-10	855362
SM4-15(2.5)9-300/100-10	855369
SM4-15(5)19-80/40-10	**689822**
SM4-15(5)19-20/200-10	**596426**
SM4-15(5)19-200/15-10	855363
SM4-15(5)19-30/100-10	855370
SM4-15(7.5)28-80/40-10	**855367**
SM4-15(7.5)28-20/200-10	**855302**
SM4-15(7.5)28-200/15-10	855364
SM4-15(7.5)28-30/100-10	855371
SM4-15(10)38-80/40-10	**682135**
SM4-15(10)38-20/200-10	**989021**
SM4-15(10)38-200/15-10	682119
SM4-15(10)38-30/100-10	855372

Recommended selections shown in bold print.

SM4-15 accessories

Description	Model number	Part number
Valve mounting bolt kit 1/4 – 20 x 2 1/4"	BK866685	866685
Valve mounting bolt kit M6 x 60 mm	BK689623M	689623
Subplate, rear port (inch)	SM4M-15-10	989027
Subplate, rear port (metric)	SM4M-15-10M	866658
Subplate, side port (inch)	SM4ME-15-10	989028
Subplate, side port (metric)	SM4ME-15-10M	866659
Subplate mounting bolt kit 1/4 – 20 x 1 1/2"	BK855992	855992
Subplate mounting bolt kit M6 x 40 mm	BK855993M	855993
Adaptor to .875 port circle (inch)	SM4A-15-M76-10	635670
Adaptor to .875 port circle (metric)	SM4A-15-M76-10M	866660
Adaptor (.875) mounting bolt kit 5/16 – 18 x 1 1/4"	BK688701	688701
Adaptor (.875) mounting bolt kit M8 x 35 mm	BK689630M	689630
Adaptor to CETOP 3 (inch)	SM4A-3-15-10	686519
Adaptor to CETOP 3 (metric)	SM4A-3-15-10M	866661
Adaptor-3 mounting bolt kit 10 – 24 x 1/2"	BK855984	855984
Adaptor-3 mounting bolt kit M5 x 12 mm	BK855985M	855985

Standard SM4-15 servo valves (cont.)

Model number	Part number
SM4-15(12.5)47-80/40-10	**683594**
SM4-15(12.5)47-20/200-10	**627657**
SM4-15(12.5)47-200/15-10	855365
SM4-15(12.5)47-30/100-10	855373
SM4-15(15)57-80/40-10	**635678**
SM4-15(15)57-20/200-10	**681347**
SM4-15(15)57-200/15-10	855973
SM4-15(15)57-30/100-10	514581

Recommended selections shown in bold print.

SM4-15 accessories (cont.)

Description	Model number	Part number
Adaptor to CETOP 5 (inch)	SM4A-5-15-10	686521
Adaptor to CETOP 5 (metric)	SM4A-5-15-10M	866662
Adaptor-5 mounting bolt kit 1/4 – 20 x 3/4"	BK855986	855986
Adaptor-5 mounting bolt kit M6 x 20 mm	BK855987M	855987
Flushing valve (inch)	SM4FV-10/15-10	633231
Flushing valve (metric)	SM4FV-10/15-10M	633231
Flushing valve mounting bolt kit 1/4–20 x 1"	BK866686	866686
Flushing valve mounting bolt kit M6 x 25 mm	BK689629M	689629
Cable connector	MS3106-14S-2S	242123
Cable clamp	MS3057-6	126058

Standard SM4-20 servo valves

Model number	Part number
SM4-20(1)3.8-80/40-10	**855380**
SM4-20(1)3.8-20/200-10	**689944**
SM4-20(1)3.8-200/15-10	689984
SM4-20(1)3.8-30/100-10	855382
SM4-20(2.5)9-80/40-10	**689823**
SM4-20(2.5)9-20/200-10	**689942**
SM4-20(2.5)9-200/15-10	687270
SM4-20(2.5)9-30/100-10	689943
SM4-20(5)19-80/40-10	**689824**
SM4-20(5)19-20/200-10	**688717**
SM4-20(5)19-200/15-10	855376
SM4-20(5)19-30/100-10	855383
SM4-20(7.5)28-80/40-10	**689825**
SM4-20(7.5)28-20/200-10	**855386**
SM4-20(7.5)28-200/15-10	855377
SM4-20(7.5)28-30/100-10	855384
SM4-20(10)38-80/40-10	**688706**
SM4-20(10)38-20/200-10	**514587**
SM4-20(10)38-200/15-10	684968
SM4-20(10)38-30/100-10	687242
SM4-20(12.5)47-80/40-10	**855381**
SM4-20(12.5)47-20/200-10	**855387**
SM4-20(12.5)47-200/15-10	855378
SM4-20(12.5)47-30/100-10	855385
SM4-20(15)57-80/40-10	**682121**
SM4-20(15)57-20/200-10	**683024**
SM4-20(15)57-200/15-10	855379
SM4-20(15)57-30/100-10	687243
SM4-20(20)76-80/40-10	**855978**
SM4-20(20)76-20/200-10	**855980**
SM4-20(20)76-200/15-10	855976
SM4-20(20)76-30/100-10	855979

Recommended selections shown in bold print.

SM4-20 accessories

Description	Model number	Part number
Valve mounting bolt kit 5/16 – 18 x 2"	BK866687	866687
Valve mounting bolt kit M8 x 50 mm	BK866690M	866690
Subplate, rear port (inch)	SM4M-20-10	682997
Subplate, rear port (metric)	SM4M-20-10M	866664
Subplate, side port (inch)	SM4ME-20-10	682998
Subplate, side port (metric)	SM4ME-20-10M	866665
Subplate mounting bolt kit 1/4 – 20 x 1 1/2"	BK855992	855992
Subplate mounting bolt kit M6 x 40 mm	BK855993M	855993
Adaptor to CETOP 5 (inch)	SM4A-5-20-10	686619
Adaptor to CETOP 5 (metric)	SM4A-5-20-10M	866666
Adaptor-5 mounting bolt kit 1/4 – 20 x 3/4"	BK855986	855986
Adaptor-5 mounting bolt kit M6 x 20 mm	BK855987M	855987
Flushing valve (inch)	SM4FV-20-10	682999
Flushing valve (metric)	SM4FV-20-10M	682999
Flushing valve mounting bolt kit 5/16–18 x 1 1/4"	BK688701	688701
Flushing valve mounting bolt kit M8 x 35 mm	BK689630M	689630
Filter module (inch)	SM4FM-20-10	686501
Filter module mounting bolt kit 5/16 – 18 x 2 3/4"	BK855421	855421
Filter module mounting bolt kit M8 x 70 mm	BK689624M	689624
Cable connector	MS3106-14S-2S	242123
Cable clamp	MS3057-6	126058

Standard SM4-30 servo valves

Standard response

Model number	Part number
SM4-30(30)113-80/65-20	**627621**
SM4-30(15)57-80/65-20	684988
SM4-30(25)95-80/65-20	681349

High response

Model number	Part number
SM4-30(30)113-80/65-21	**855349**
SM4-30(15)57-80/65-21	855337

Robot standard response

Model number	Part number
SM4-30(25)95-80/65-20	681346

Robot high response

Model number	Part number
SM4-30(25)95-80/65-21	593591

Recommended selections shown in bold print.

SM4-30 accessories

Description	Model number	Part number
Valve mounting bolt kit 1/4 – 20 x 1 1/4"	BK866688	866688
Valve mounting bolt kit M6 x 35 mm	BK689626M	689626
Subplate, rear port (inch)	SM4M-30-20	627617
Subplate, rear port (metric)	SM4M-30-20M	866669
Subplate, side port (inch)	SM4ME-30-20	627616
Subplate, side port (metric)	SM4ME-30-20M	866670
Subplate mounting bolt kit 1/4 – 20 x 2 1/4"	BK866685	866685
Subplate mounting bolt kit M6 x 35 mm	BK689623M	689623
Adaptator to CETOP 5 (inch)	SM4A-5-30-20	686621
Adaptator to CETOP 5 (metric)	SM4A-5-30-20M	866671
Adaptator-5 mounting bolt kit 1/4 – 20 x 3/4"	BK855986	855986
Adaptator-5 mounting bolt kit M6 x 20 mm	BK855987M	855987
Flushing valve (inch)	SM4FV-30-20	633230
Flushing valve (metric)	SM4FV-30-20M	633230
Flushing valve mounting bolt kit 1/4–20 x 1 1/4"	BK866688	866688
Flushing valve mounting bolt kit M6 x 35 mm	BK689626M	689626
Cable connector	MS3106-14S-2S	242123
Cable clamp	MS3057-6	126058

Standard SM4-40 servo valves

Model number	Part number
SM4-40(20)76-80/40-10	**855392**
SM4-40(20)76-20/200-10	**689785**
SM4-40(20)76-200/15-10	855388
SM4-40(20)76-30/100-10	855394
SM4-40(25)95-80/40-10	**689826**
SM4-40(25)95-20/200-10	**681348**
SM4-40(25)95-200/15-10	855389
SM4-40(25)95-30/100-10	855395
SM4-40(30)113-80/40-10	**686523**
SM4-40(30)113-20/200-10	**684992**
SM4-40(30)113-200/15-10	855390
SM4-40(30)113-30/100-10	855396
SM4-40(35)132-80/40-10	**689827**
SM4-40(35)132-20/200-10	**855399**
SM4-40(35)132-200/15-10	855391
SM4-40(35)132-30/100-10	855397
SM4-40(40)151-80/40-10	**855393**
SM4-40(40)151-20/200-10	**855983**
SM4-40(40)151-200/15-10	689951
SM4-40(40)151-30/100-10	855398

Recommended selections shown in bold print.

SM4-40 accessories

Description	Model number	Part number
Valve mounting bolt kit 5/16 – 18 x 3"	BK866689	866689
Valve mounting bolt kit M8 x 80 mm	BK689628M	689628
Subplate, rear port (inch)	SM4M-40-10	989025
Subplate, rear port (metric)	SM4M-40-10M	866673
Subplate, side port (inch)	SM4ME-40-10	989026
Subplate, side port (metric)	SM4ME-40-10M	866674
Subplate mounting bolt kit 1/4 – 20 x 2 1/4"	BK866685	866685
Subplate mounting bolt kit M6 x 60 mm	BK689623M	689623
Adaptator to CETOP 8 (inch)	SM4A-8-40-10	686623
Adaptator to CETOP 8 (metric)	SM4A-8-40-10M	866675
Adaptator-8 mounting bolt kit 1/2 – 13 x 2 1/4"	BK855990	855990
Adaptator-8 mounting bolt kit M12 x 60 mm	BK855991M	855991
Flushing valve (inch)	SM4FV-40-10	635681
Flushing valve (metric)	SM4FV-40-10M	635681
Flushing valve mounting bolt kit 5/16–18 x 1 1/4"	BK688701	688701
Flushing valve mounting bolt kit M8 x 35 mm	BK689630M	689630
Cable connector	MS3106-14S-2S	242123
Cable clamp	MS3057-6	126058

Index

S

T

V

V

Z